Project AIR FORCE

PATTERNS IN CHINA'S USE OF FORCE

EVIDENCE FROM HISTORY AND DOCTRINAL WRITINGS

MARK BURLES
ABRAM N. SHULSKY

Prepared for the
UNITED STATES AIR FORCE

RAND

The research reported here was sponsored by the United States Air Force under Contract F49642-96-C-0001. Further information may be obtained from the Strategic Planning Division, Directorate of Plans, Hq USAF.

Library of Congress Cataloging-in-Publication Data

Burles, Mark, 1970-
    Patterns in China's use of force : evidence from history and
doctrinal writings / Mark Burles and Abram N. Shulsky.
      p. cm.
    " MR-1160-AF."
    Includes bibliographical references.
    ISBN 0-8330-2804-9
    1. China—Defenses. 2. China—History, Military—1949-
    3. China—Military relations—United States. 4. United
    States—Military relations—China. 5. Strategy. I. Shulsky,
    Abram N. II. Title.

UA835 .B87 1999
355'00951—dc21   99-088091

RAND is a nonprofit institution that helps improve policy and decisionmaking through research and analysis. RAND® is a registered trademark. RAND's publications do not necessarily reflect the opinions or policies of its research sponsors.

Published 2000 by RAND
1700 Main Street, P.O. Box 2138, Santa Monica, CA 90407-2138
1333 H St., N.W., Washington, D.C. 20005-4707
RAND URL: http://www.rand.org/
To order RAND documents or to obtain additional information,
contact Distribution Services: Telephone: (310) 451-7002;
Fax: (310) 451-6915; Internet: order@rand.org

China is emerging as a major global and regional player that will impact U.S. foreign policy well into the 21st century. A better understanding of China's interests as well as economic and military capabilities will assist in crisis prevention and war avoidance.

This study examines the characteristic ways in which China might use force to protect or advance its interests. It looks at the record of Chinese use of force during the past 50 years, as well as at Chinese doctrinal writings concerning future conflict to understand what particular characteristics future Chinese uses of force might be expected to display.

This research was conducted within the Strategy and Doctrine Program of Project AIR FORCE, as part of a larger project entitled "Chinese Defense Modernization and the USAF," under the sponsorship of the Deputy Chief of Staff for Air and Space Operations (AF/XO), and the Commander, Pacific Air Forces (PACAF/CC). Comments are welcomed and may be addressed to the project leader, Dr. Zalmay Khalilzad, or to the authors.

PROJECT AIR FORCE

Project AIR FORCE, a division of RAND, is the Air Force federally funded research and development center (FFRDC) for studies and analyses. It provides the Air Force with independent analyses of policy alternatives affecting the development, employment, combat readiness, and support of current and future aerospace forces. Research is performed in four programs: Aerospace Force Development; Manpower, Personnel, and Training; Resource Management; and Strategy and Doctrine.

CONTENTS

The way the People's Republic of China (PRC) has used force in crisis and conflict situations has often surprised and perplexed the United States, as well as other countries. Some insight into the likely characteristics of any future Chinese use of force is important for U.S. policymakers who will have to deal with possible Chinese actions in future crises and conflicts. Without attempting to predict whether China will resort to the use of force in any given situation, it may be possible to gain some insight into ways in which the Chinese, *if* they decide to use force, may be guided by a strategic understanding that differs from that with which we are familiar and that relies on concepts other than those—such as deterrence and coercion—in terms of which we understand the use of force. While the focus is on the question of *how* the Chinese might use force, the answer has implications for the question of whether, in any given situation, they are likely to do so. In particular, the Chinese appear to believe that they possess tactics and methods that make it feasible for them to use force even when the overall military balance is very unfavorable to them, i.e., in situations in which their use of force might otherwise have been thought to be very unlikely.

This report exploits two avenues for approaching this issue: the 50-year historical record of PRC actions and the evolution of their doctrinal writings on preparing for, and fighting, a future war. The historical record shows a pattern of using force in a conflict to achieve surprise and thus administer a strong psychological or political shock to the adversary. By upsetting the adversary's strategy and expectations, China hopes to force it to make a radical reevaluation of its goals and to acquiesce in a new *status quo* that is much more favorable to China. In some cases, e.g., the 1962 Chinese invasion of contested territory along the Sino-Indian border, the entire purpose of the operation appears to have been to administer a psychological shock; instead of holding on to the captured territory as a "bargaining chip," which might have seemed to be the more obvious strategy, the Chinese withdrew from it unilaterally.

In addition, the Chinese have used force to create a sense of crisis, even though it was militarily inferior to its potential opponent. It is often claimed that, since *weiji*, the Chinese word for *crisis*, is composed of two characters, which can be translated as *danger* and *opportunity*, respectively, it does not have an entirely negative connotation. This is apparently apocryphal[1]; nevertheless, it does appear that, for the Chinese leadership, a crisis is not necessarily a negative phenomenon: It may provide an opportunity for making gains that would otherwise not be achievable. The view that crises are invariably bad—i.e., that they offer "danger" but not "opportunity"—is the understandable perspective of a *status quo* power, such as the United States, that seeks primarily to avoid war, especially nuclear war. Thus, the creation of a crisis may be a way to probe an adversary's intentions, to cause difficulties between him and his allies, or to weaken his resolve and the domestic political support for his policies.

When one turns from the historical record to Chinese doctrinal writings on national military strategy, one sees an evolution from an earlier strategy of "people's war" to "local war under high-tech conditions." "People's war" was a reactive strategy to deal with a "worst case" scenario, i.e., a massive invasion by a militarily superior power. It was designed to address this threat through a prolonged conflict of attrition. Obviously, the PRC never in fact fought such a war after 1949. However, the concept presumably guided Chinese preparations for a future war, including such issues as the structure of its armed forces and the attempt, via the "third front," to build secure industrial and logistics bases in remote (from the likely invasion routes) and mountainous areas of China.

By the mid-1980s, however, the Chinese political leadership had concluded that the risk of a major invasion had passed, and China's

---

[1] Or so the authors understand from a native speaker.

People's Liberation Army (PLA) was redirected toward preparations for a smaller-scale "local war." Unlike "people's war," the military demands of a local war place a premium on the PLA's ability to gain the initiative at the earliest stage of the conflict, possibly through preemption. China's military strategy, therefore, is much more suited now to diplomatic strategies that call for the opportunistic or demonstrative use of force to further Chinese foreign policy interests.[2]

Examination of recent examples of local war, of which the most prominent is the Gulf War of 1991, quickly shows that the PLA, although stronger than the armed forces of many of the neighboring countries, is ill-equipped to fight such a war. It lacks the high-technology platforms and weapons that such a war requires, and its troops are insufficiently trained. The PLA's large size and light infantry emphasis, an inheritance from the "people's war" concept, are hindrances in this regard. Thus, the major problem for the PLA becomes, not so much the prosecution of a high-tech campaign, but the disruption of such an action on the part of a potential adversary. In particular, the PLA must seek to understand the vulnerabilities of the U.S. armed forces that it might be able to exploit.

Chinese authors have identified several vulnerabilities that they believe China may be able to exploit: long logistics lines to the western Pacific; dependence on sophisticated, but potentially fragile and unreliable, information technology; difficulties of conducting naval operations (e.g., antisubmarine warfare and mine clearing) in shallow waters; and casualty intolerance. Conversely, Chinese use of dispersal, camouflage, and deception can degrade U.S. offensive strengths in the area of precision strike.

The disturbing conclusion of this analysis is that China, despite an awareness of its relative weakness, might nevertheless be willing to use force against the United States or in a way that runs a major risk of U.S. involvement. In using force in this way, China would be primarily seeking to achieve a political effect. One can easily imagine circumstances in which the Chinese might believe such a political effect could be obtained. The most obvious cases would deal with

---

[2]This is not to say that such uses of force did not occur before the mid-1980s; the point is that the *doctrine* changed at that time.

Taiwan and would seek to exploit the ambiguities in the U.S. commitment to defend Taiwan, especially under circumstances in which a Taiwanese action (perhaps even a less-provocative action than a unilateral declaration of independence) could be said to have precipitated the crisis. In such a case, a Chinese use of force could be directed toward affecting U.S. policy and driving a wedge between Taiwan and the United States.

Force could also be used to influence the political situation on Taiwan. By raising fears of a major military action, China might hope to exacerbate tensions on Taiwan between those willing to run major risks in the name of eventual independence and those who, whatever they thought about independence in the abstract, did not wish to run risks to change the *status quo*. Other possibilities for Chinese use of force are more remote. One could imagine Chinese military actions in the South China Sea in support of its territorial claims. While this could easily involve minor incidents (such as naval skirmishes or attacks on islands held by other claimants), it is more difficult to imagine how a major clash between the United States and China might result.

In any conflict or potential conflict with the United States, China, understanding that it is the generally weaker party, would have to look for asymmetric strategies that would provide leverage against the United States and exploit U.S. vulnerabilities, while preventing the United States from bringing its superior force to bear. China would seek to create a *fait accompli*, thereby forcing the United States, if it wished to reinstate the *status quo ante*, to escalate the level of tension and violence. China would then count on the pressure of "world public opinion," the general disinclination to see profitable economic relationships disrupted, and U.S. public opinion to constrain the United States in this situation.

In support of this goal, China would seek to suppress the U.S. ability to project substantial military resources into the theater of conflict for a limited time, either by information warfare attacks or by missile attacks (or the threat of them) on ports, airfields, transit points, bases, or other key facilities in the western Pacific.

China could seek to cause U.S. casualties to shock U.S. public opinion. The articles that discuss this issue frequently point to the U.S. experiences in such places as Somalia and Lebanon, where U.S.

forces did withdraw after suffering unexpected casualties. A major psychological shock would be sought: China could seek to do something for which public opinion would not be at all prepared. Finally, China could seek to exploit the fact that the United States was dealing with a crisis elsewhere in world; for example, it could act against Taiwan at a time when United States had made major deployments to the Persian Gulf.

Needless to say, these are extremely risky strategies, since, like the Japanese attack on Pearl Harbor, they could easily lead to a major U.S. response that, in time, would prove overwhelming. The Chinese leadership could not embark on them unless it was confident that it could assess the likelihood of such a U.S. reaction. Among other things, China would have to take into account the U.S. stake in the conflict and could not reasonably expect that losses of the magnitude that led to the U.S. pullout from Somalia would have the same effect in East Asia.

Similarly, the Chinese leadership would have to assess the risks of widening the conflict if it, e.g., threatened to attack U.S. bases or port or other facilities to which U.S. forces had access located on the territory of third-party countries. While Chinese attacks, or threats of attack, against such facilities could delay U.S. deployments via those facilities and/or complicate U.S. access to them, the result could also be to increase the hostility of third-party countries; China could find that its action had served mainly to strengthen the cohesion and will of a coalition directed against it.

The PRC's historical record shows that its leaders have been willing to take risks of this sort and have, in fact, been quite successful in assessing them.[3] In the historical cases discussed in Chapter Two, the PRC was able to modulate the risk so as to avoid a massive reaction from its stronger adversaries (both the United States and the former Soviet Union). In many cases, China tried to create the appearance of engaging in bolder actions than it was really undertaking; during the Taiwan Strait crises of 1954–1955 and 1958, for instance, China apparently relied on the United States to understand

---

[3]While this is particularly true of Mao Zedong, even the pragmatic Deng Xiaoping was willing to run the risk of a serious Soviet reaction in 1979 to "teach a lesson" to Vietnam.

that, despite its rhetoric about "liberating Taiwan," it was in fact posing no such threat.[4]

China's past success in assessing and modulating the risk of a massive reaction by its adversaries may give it confidence that it will be able to do so in the future as well. In addition, China may feel, now and in the future, that it can afford to accept greater risks. Many of the past uses of force occurred when China either was not a nuclear power or did not have a secure nuclear second-strike capability. The possession of strategic nuclear weapons may enable the Chinese leadership to run risks that it otherwise could not. [5]

On the other hand, China ran its past risks when it could, to some extent, count on support of one superpower against the other. Even in 1969, before the Sino-U.S. *rapprochement*, China gained some benefit from the superpower rivalry; the Soviet Union had to fear that the United States would exploit any opportunities created by a major war between it and China. The U.S. status as sole superpower reduces China's maneuvering room, which helps explain the Chinese preference for multipolarity. While the Chinese typically argue that the international system is evolving in the direction of multipolarity, it is also clear that this tendency has not progressed very far. In this respect, then, China faces a more difficult environment in which to run risks. In addition, after decades of economic development, China has more to lose if it underestimates the risks it is running than it did in the earlier years of the PRC.

---

[4]To some extent, this undercuts the purpose of the operation, which is to place psychological pressure on the adversary. However, the audiences for the two messages are slightly different: One can hope to convince the adversary's public (and even a part of its political elite) that the risk of war is increasing even if military and intelligence specialists understand that there is less to the threat than meets the eye.

[5]To some extent, the "people's war" strategy, which postulated China's invulnerability to ultimate capture and defeat, served the same purpose.

ACKNOWLEDGMENTS

The authors wish to thank the project's action officers, Majors Stephen Cunico and Milton Johnson of the Directorate of Air and Space Operations, Headquarters, U.S. Air Force, and James Hertsch of the National Air Intelligence Center, for their help and encouragement. Thomas J. Christensen and David M. Finkelstein reviewed an earlier version of this manuscript and provided many helpful comments and suggestions, for which the authors are very grateful.

The authors also wish to thank RAND colleagues Roger Cliff and James Mulvenon for their criticism and suggestions; research assistant Andrew Mok; editor Phyllis Gilmore; and, in particular, Luetta Pope and Grace Young, for their able secretarial assistance.

ABBREVIATIONS

ASEAN	Association of Southeast Asian Nations
C^3	Command, control and communications
C^3I	Command, control, communications and intelligence
CCP	Chinese Communist Party
CMC	Central Military Commission (of the PRC)
FBIS	Foreign Broadcasting Information Service
GMD	Guomindang (or Kuomintang, ruling party of the Republic of China)
JPRS	Joint Publication Research Service
NCNA	New China News Agency (official PRC news service)
PLA	People's Liberation Army
PLAAF	People's Liberation Army Air Force
PLAN	People's Liberation Army Navy
PRC	People's Republic of China

Chapter One

INTRODUCTION

This report examines the particular characteristics that a future Chinese use of force might exhibit. It does not attempt to predict Chinese behavior in any given situation, since such behavior will be determined by the specific political and military conditions at the time. In particular, it does not argue that China is eager to use force or is singularly likely to do so. It does, however, attempt to provide some insight into ways in which the Chinese, *if* they decide to use force, may be guided by a strategic understanding that differs from the one with which we are familiar. By explicating for policymakers some of these characteristic differences, the report will contribute to a better understanding of possible Chinese actions in future crises and conflicts. While the focus is on the question of *how* the Chinese might use force, the answer has implications for the question of whether, in any given situation, they are likely to do so. In particular, as will be argued, the Chinese appear to believe that—given the tactics and methods they would employ—it is feasible for them to use force even when the overall military balance is very unfavorable to them, i.e., in situations in which their use of force might otherwise have been thought to be very unlikely.

The report is based primarily on the historical record of the People's Republic of China (PRC) since 1949 and on the published doctrinal and policy statements of Chinese officials and commentators.[1]

---

[1]The historical experience of imperial China with respect to the use of force is discussed at length in another report prepared in connection with this project (Swaine and Tellis, forthcoming).

The PRC historical record, while rich in examples of the use of force, is limited in the sense that it deals with a period during which China was relatively weak compared to the main threats it feared, i.e., the United States and the Soviet Union. Thus, most of the available examples involve the use of force against a stronger power or against a client of a stronger power. The major exceptions (aside from the somewhat problematic case of the border war with India in 1962) involve the use of force to gain control of islands in the South China Sea and the "lesson" China attempted to teach Vietnam in 1979. However, with respect to the uses of force that are of most concern, i.e., those that could lead to a Sino-U.S. crisis, this situation of Chinese overall relative weakness will most likely continue for the next several decades. The important point to be recognized is that this relative weakness *cannot* necessarily be relied upon to deter China from using force. This historical record is discussed in Chapter Two.

Chapter Three examines the evolution of China's military strategy during the half-century of the PRC's existence. During its early years, PRC military strategy focused on the "worst case" scenario (i.e., all-out attack and invasion by the United States or, later, the Soviet Union), but the PRC never in fact had to implement this strategy; since 1985, military strategy has focused on the types of "local war" in which the PRC had in fact been involved. The political-military content of the "local war" concept is defined by the constraints imposed by current international circumstances. Fundamental to these constraints is what the Chinese see as the primacy of economic development, especially in East Asia; to the extent that most nations are concerned with and focused on economic development, their tolerance for long, disruptive wars is limited. These political constraints heighten the importance of new technologies (such as precision strike) that make it possible to attain decisive military results much more rapidly than in the past.

Chapter Four then examines, in the light of the historical record and of the "local war" concept, the political circumstances under which China might use force in the future and the strategies that might guide that use of force.

From these political-military considerations, we then turn to the more technical aspects of how China might use force in the future. Chapter Five discusses possible Chinese military options based on

Chinese strategists' view of the nature of conflict in the post–Cold War world and the strengths and weaknesses of the People's Liberation Army (PLA). It is based on an interpretation of the "local war under high-tech conditions" military strategy, which represents China's view of the most effective way to organize and use its military forces to protect or pursue its national interests in the current international environment.[2]   China's military strategy lends valuable insight into Chinese strategists' views of their security environment, the nature of future military conflict, and the kinds of military capabilities they will need to be effective in those conflicts.

Chapter Six then looks at how Chinese strategists have applied these concepts to the case of a possible future conflict with the U.S. Chinese military strategists appear to understand the overwhelming superiority the United States enjoys in almost every aspect of military power. Nonetheless, they do identify a number of U.S. vulnerabilities that the PLA could exploit should conflict occur.

Chapters Five and Six rely on the writings of Chinese officials and observers. A number of important caveats regarding the nature of the sources used for this report warrant mentioning at this point. Although the newspapers and periodicals cited are, generally, official publications of the Chinese government, they do not necessarily reflect official Chinese policy on any given topic. While it is not likely that an article could be published that directly contradicted an established policy, PRC publications do contain a range of opinions that reflect real debates within China over security policy. In addition, some articles may represent special pleading on the part of individual military services, whose interests lie in convincing China's military leaders of the particular effectiveness of a certain kind of weapon system or strategic concept.[3]   To ensure some degree of

---

[2]Then–Central Military Commission (CMC) Vice-Chairman Liu Huaqing articulated this definition of National Military Strategy in a 1993 *Jiefangjun Bao* article (Liu Huaqing, 1993).

[3]Such special pleading can be seen in a Chinese document that is discussed and excerpted in Munro (1994). The report Munro examined clearly promotes parochial naval interests in its description of likely future security threats confronting the PRC. For example, it argues that the most effective military strategy to defeat a hostile U.S. force on the Korean Peninsula would be through an amphibious assault across the Yellow Sea, rather than through the deployment of ground forces across the Yalu. To be able to conduct such difficult naval operations, the report calls for China to have two carrier task forces by 2001.

authenticity, most quotes selected are from high-level military leaders or from personnel from China's primary military academies.[4]

The articles available in open sources tend, not surprisingly, to deal with technical military issues in rather general terms. In many cases, articles that discuss developments falling under the heading of the "revolution in military affairs" will sound familiar to those who have followed the similar literature published in the United States. Thus, it is difficult to know how much detailed investigation by Chinese researchers lies behind any of the general statements contained in the articles. In particular, one cannot be sure whether a given statement (to the effect, for example, that the PLA could target key U.S. logistics nodes) represents the result of a detailed Chinese analysis of a military scenario, a suggestion for how Chinese planners might approach their task in the future, or simply a reflection of concerns expressed in the United States about possible "asymmetric strategies" that an adversary could adopt.

Beijing is reluctant to make explicit policy pronouncements on what it considers to be sensitive issues related to national military strategy. Most articles offer only vague descriptions of the kinds of security threats or issues that the PRC believes hold the greatest potential for provoking military conflict. Finally, the articles are largely drawn from translations provided by the Foreign Broadcasting Information Service (FBIS). Some question remains as to how comprehensively this covers the range of literature on this topic.

The report concludes (in Chapter Seven) with a discussion of the implications for the U.S. armed forces. The appendix presents a short note on the debate over Chinese "strategic culture" and suggests that the characteristic ways the PRC has used force are consonant with classical Chinese military thought (as represented by Sun Zi's *The Art of War*), especially when viewed in distinction to a canonical Western military thinker, such as Carl von Clausewitz.

---

[4]China has two primary military academies, the National Defense University and the Academy of Military Science, whose functions vary somewhat. The university's role is to inculcate PLA officers in the principles of China's current military doctrine. The academy's role is more forward looking. It identifies and examines future trends in warfare and reports to the CMC on its findings.

Chapter Two

PATTERNS IN THE PRC'S USE OF FORCE

This chapter briefly reviews the PRC historical record to glean lessons for understanding potential future PRC political-military behavior. Most of the examples come from the Mao period; however, the use of force in the post-Mao period, although much rarer, exhibits some of the same characteristics. After examining some typical patterns in the PRC's use of force, we will relate them to the overall context within which Chinese national strategy has developed and to current Chinese doctrinal discussions of these issues.

CHARACTERISTICS OF CHINESE USE OF FORCE IN AN ACTUAL CONFLICT

Surprise

A key characteristic of Chinese use of force in actual conflict has been the importance of the element of surprise. While we often tend to think of surprise as being equivalent to a "bolt from the blue," i.e., a no-warning attack on a force that had no idea an attack was coming, this is not necessarily the case. The key point is not the *subjective* sense of surprise experienced by the targets of the attack (and certainly not whether they had enough evidence of an impending attack that they should not have been surprised), but rather whether or not they had in fact made adequate preparations for it. By that standard, the Chinese have often been successful at achieving surprise.[1]

---

[1]Surprise evidently requires some degree of denial and deception; however, successful deception typically involves some (and, often, a great deal of) self-deception on the part of the target. Hence, it is difficult in any given case to decide how much of the

In the Korean War, for example, the Chinese attack in force at the end of November 1950 achieved tactical surprise, with devastating effects on the U.S. forces. Although it was known that Chinese forces had crossed the Yalu River into Korea and had indeed made contact with U.S. forces several weeks earlier, the actual attack found the U.S. forces badly positioned and unprepared for an enemy that differed greatly from the Soviet-style North Korean army they had just soundly defeated.[2]

While much of the difficulty the U.S. forces found themselves in can be attributed to an "intelligence failure" on the part of the United States, one can identify several courses of action the Chinese took that facilitated that failure and heightened the effects of the successful surprise. These include

- breaking off contact after the initial confrontation between Chinese and U.S. troops in late October–early November

- releasing U.S. troops taken captive in those encounters

- exploiting General MacArthur's impatience to reach the Yalu River before it froze.[3]

The net result of the first two steps was to reinforce U.S. inclinations to believe that the earlier Chinese warnings against crossing the 38th parallel were bluffs, that the Chinese were not anxious to fight the United States, and that a rapid advance to the Yalu could end the war before China could do very much about it. Documents that have recently become available support the idea that, to some extent at least, the Chinese may have been deliberately trying to encourage

---

"credit" for a successful surprise should go to the surpriser and how much to the victim.

[2]For the nature of the surprise the U.S. forces suffered in Korea, see, in particular, Cohen and Gooch (1991), pp. 175–182. They argue that this aspect of the U.S. "intelligence failure" in Korea—the failure to anticipate that Chinese "human wave" infantry attacks would require different tactics to defeat than the motorized North Korean army—was the more important failure, as opposed to a presumed failure to anticipate Chinese intervention on a massive scale. See also pp. 192–194 for MacArthur's view of the significance of the Chinese lack of air cover for their ground forces.

[3]The winter freeze would hamper U.S. efforts to cut off Chinese reinforcements, since destroying the bridges over the Yalu would no longer be an effective tactic once the river had frozen.

what they saw as MacArthur's overconfidence and that they took steps (such as releasing the Americans captured in the first skirmishes) to encourage it:

> On November 18, 1950, one week before MacArthur's offensive, Mao sent a telegram to [Marshall] Peng [Dehuai] celebrating the American misperception of China's troop strength. Mao knew that MacArthur falsely believed Chinese forces in Korea to consist of only 60,000 or 70,000 troops, when actually there were at least 260,000. Mao told Peng that this was to China's advantage and would assist Chinese forces in destroying "tens of thousands" of enemy troops. In the same telegram, Mao instructed Peng to release prisoners of war. (Christensen, 1996b, p. 171.)[4]

Marshall Peng later claimed that the release of United Nations prisoners was designed to encourage MacArthur's further advance northward. (Christensen, 1996b, p. 171.) According to Peng,

> First, though we achieved success in the first offensive operation, the enemy's main force remained intact. With the main body of the CPV [Chinese People's Volunteers] unexposed, it was expected that the enemy would continue to stage an offensive. Second, the enemy had boasted the ability of its airforce [sic] to cut off our communication and food supply. This gave us an opportunity to deceive the enemy about our intention. By releasing some POWs [prisoners of war], we could give the enemy the impression that we are in short supply and are retreating. Thirdly, the enemy is equipped with air and tank cover, so it would be difficult for us to wipe out the retreating enemy on foot. (Hao and Zhai, 1990, pp. 113–114.)[5]

The result was that the Chinese achieved effective tactical surprise despite the fact that the United States was, generally speaking, aware of the presence of Chinese troops in Korea. The Chinese successes in

---

[4]Christensen cites a telegram dated November 18, 1950, to Peng Dehuai and others concerning the release of prisoners of war (found in *Jianguo Yilai Mao Zedong Wengao*, Vol. 1, p. 672).

[5]Hao and Zhai cite Yao Xu, *From Yalu River to Panmunjon*, Beijing: People's Press, 1985, pp. 39–40, as their source.

the initial engagements illustrate the doctrine of the classic Chinese strategist, Sun Zi (1:17–18, 23, 26–27[6]):

> All warfare is based on deception.
>
> Therefore, when capable, feign incapacity; when active, inactivity.
>
> . . .
>
> Pretend inferiority and encourage his arrogance. . . .
>
> Attack where he is unprepared; sally out when he does not expect you.
>
> These are the strategist's keys to victory. It is not possible to discuss them beforehand.

The importance of surprise is also illustrated in the Chinese operations against India in 1962. Overlapping Indian and Chinese territorial claims, and Indian insistence on establishing military outposts in disputed border areas (India's "forward policy"), led to fighting between the two countries in 1962. On October 20, the Chinese launched major attacks, which succeeded immediately. After a short pause, the Chinese renewed their attacks in mid-November (in one case, an Indian offensive action planned for November 14 failed, leading to a successful counterattack), and by November 20, no organized Indian units remained in the entire disputed territory.

Throughout the months leading up to the border war of October tp November 1962, the Indians held a mistaken view of China's willingness to resist their "forward policy." To some extent, this misunderstanding derived from India's belief that China had been so weakened by the disastrous policies of the "Great Leap Forward" that it had no choice but to acquiesce. But it was also fed by specific Chinese acts, both verbal and on the ground.

During the months before the conflict, China issued a series of warnings, both diplomatic notes and public statements, attacking the Indian "forward policy." Nevertheless, the tone was often relatively restrained. For example, the Chinese, in a diplomatic note dated June 2, 1962, after accusing the Indians of seeking "to provoke

---

[6]In this and all subsequent references to Sun Zi's *The Art of War*, the number before the colon refers to the chapter, the number(s) after to the verse(s). We have used the Griffith (1963) translation.

bloody conflicts, occupy China's territory and change the status quo of the boundary regardless of consequences," nevertheless concluded as follows:

> The Chinese government consistently stands for a peaceful settlement of the Sino-Indian boundary question through negotiations. Even now when the Sino-Indian border situation has become so tense owing to Indian aggression and provocation, the door for negotiations is still open so far as the Chinese side concerned [sic]. However, China will never submit before any threat of force.[7]

Other notes warned that "India will be held responsible for all the consequences" arising from its intrusions; but this was coupled with an assertion that the Chinese government strove "to avoid clashes with intruding Indian troops."[8] The Indians may have regarded this verbal posture as not particularly threatening; it seemed to imply only that the Chinese would fight back if attacked. Thus, the Indians may have believed that, as long as their forces did not initiate conflict with the Chinese, the "forward policy" could continue as a form of shadow boxing, with each side maneuvering its patrols to gain positional advantage but without actually coming to blows.

Perhaps more importantly, Chinese behavior was restrained as well. The Indian "forward policy" inevitably led to situations in which units of the two armies confronted each other in close proximity. In two instances, in May and July 1962, Chinese troops adopted threatening postures with respect to newly implanted Indian posts; in the latter case, they surrounded an Indian post and blocked an attempt to resupply it by land, forcing it to rely on airdrops. Nevertheless, in both cases, the Chinese did not follow through on their implicit threats and refrained from actually attacking the posts. This served to confirm, for the Indians, the wisdom of the "forward policy": If the Indians were resolute, the Chinese would not use force to interfere with their strategy of creating a network of new posts. (Maxwell, 1970, p. 239.) While there is no reason to believe that these Chinese statements and actions were designed to deceive the Indians in the interest of achieving tactical surprise, they certainly contributed to it.

---

[7]New China News Agency—English, June 9, 1962.

[8]Chinese government memorandum of July 8, 1962, handed to the Indian chargé d'affaires in Beijing, New China News Agency—English, July 8, 1962.

China's "pause" between its initial victories in October and its all-out offensive in mid-November may also have served to give the Indians an unfounded confidence in their assessment of Chinese intentions.[9] During the pause, Chinese infantry used a side trail to infiltrate deep into the disputed territory, outflanking the new Indian defensive position established after the initial reverse. But if, for the Chinese, the pause presented the opportunity to regroup and infiltrate troops behind the main Indian positions, the Indians appeared to interpret it as a sign that the Chinese would not, or could not, pursue their attack any further. At the same time, the Indians held an exaggerated view of their own strength, believing that they were now able to resist any Chinese attack.[10]

Psychological-Political Shock

Surprise not only confers a military advantage but also heightens the psychological or political shock value of any initial military success. The latter effect may have important payoffs: The adversary may become disheartened and defeatist as a result of the unexpected reverse he has suffered and may be induced to reduce his war aims. The importance of achieving such a psychological-political shock is magnified in the case of a military force that has limited force-projection and sustainment capabilities.

In the course of their intervention in Korea, the Chinese presumably meant to drive the U.S. forces off the peninsula altogether.[11] Poor Chinese logistics and the lack of effective air cover meant that, the

---

[9]This is similar to what occurred in Korea in November 1950.

[10]For example, on November 12, Home Minister Lal Bahadur Shastri claimed that "India was now strong enough to repulse the Chinese attackers and was building its military might to drive the invaders from Indian soil." (Maxwell, 1970, p. 387.)

[11]Christensen (1996b) notes on p. 166 that,

> Since what Mao feared most was a local military deadlock in Korea while China remained vulnerable to American bombing, he tried to eliminate the possibility of such a deadlock by seeking total destruction of American forces in Korea. Mao consistently rejected any notion of deterrence or local military compromise from October 1950 through January 1951. . . . [H]is minimum defensive needs led him to reject the notion of a buffer anywhere in Korea, north or south of the 38th parallel. Maintaining such a defensive line would have carried many of the costs of passive border defense in Manchuria and would also sustain the risk of a future two-front war with the United States.

further north the initial engagement between the Chinese and U.S. troops occurred, the more advantageous for the Chinese. Hence, China's best hope of success was to inflict such a large and demoralizing defeat on the U.S. forces in the initial encounters (when they were relatively far north) as to maximize the possibility that the political shock would lead to a U.S. decision to withdraw from the Korean peninsula altogether.[12] Although the eviction of U.S. forces from the peninsula might seem, in retrospect, to have been an impossibly ambitious goal, it did not seem that way at the time.[13]

Similarly, the magnitude and unexpectedness of the Chinese victories over the Indians in 1962 had (from the Chinese point of view) an advantageous effect on Indian policy. Although the Indians had expected that a Sino-Indian war, if it came, would likely become a long-drawn-out affair,[14] the shock of the Chinese victory in November led them to abandon their "forward policy" and, in effect, the territorial claims it was meant to support.[15] This was true even

---

[12]The same reasoning led Mao to prefer, in late December 1950, that a follow-on Chinese offensive against the U.S. troops *not* push them out of Seoul toward the relative safety of bases further south. (Christensen, 1996b, p. 173.)

[13]For a description of the consideration in Washington in early December 1950 of the proposition that it might be necessary to withdraw from the Korean peninsula entirely, see Kennan (1972), pp. 26–33. Particularly striking is a comment of General George Marshall (as recorded by Kennan) to the effect that "[i]t was impossible to determine at the present moment whether any line or beachhead could be held."

[14]With respect to the long-term and large-scale strategic planning that, as General Staff Director of Military Operations, he was engaged in at the time, Indian Major General D. K. Palit commented:

> Today [1991] . . . this strategic overview will seem archaic, naïve even, but we were then closer to the Second World War in time and outlook than we were to notions of international restraints and the political complexities of the 1970s and 1980s. We still regarded warfare in the absolute, decisive terms and wider strategic horizons of the 1940s. Concepts such as limited wars, superpower management of local conflicts, proxy wars and coercive diplomacy had not yet modified the perception that decision was the goal of war. *It was left to the Chinese to point the way to subtleties of contemporary political and strategic manoeuvres.* (Palit, 1991, pp. 280–281; emphasis added.)

[15]Briefly, the conflicting territorial claims are as follows: India claimed Chinese-held territory on the western part of the border (the Aksai Chin region, near Kashmir), while the Chinese claimed Indian-held territory in the east, between Burma and Bhutan (the North East Frontier Agency, lying south of the "McMahon Line" of 1914, whose legitimacy the Chinese did not recognize). The Indian "forward policy" involved placing outposts in Indian-claimed territories in the west and along the MacMahon Line. The Chinese captured the agency region in November 1962 and then withdrew from it.

though the Chinese, after achieving their victory, almost immediately withdrew from the disputed area they had invaded in northeastern India; the political effect of this withdrawal was magnified by the fact that it was announced unilaterally by the Chinese (rather than negotiated with the Indians).

In essence, rather than using the territory they had just captured as a "bargaining chip" to get the Indians to abandon their "forward policy," the Chinese relied on the psychological and political shock of their victory to achieve the same end. There were, of course, other motives for the unilateral withdrawal; most obviously, the Chinese had to be concerned by the difficulty that the approaching winter would create for them in sustaining their troops in the foothills on the southern (Indian) side of the Himalayas.[16] Nevertheless, the strategy adopted seems consistent with oft-noted Chinese notions of "punishing" or "teaching a lesson": By sharply undercutting Indian self-confidence, the Chinese achieved a more rapid and decisive change in policy than would likely have emerged from a long-drawn-out negotiation in which the Chinese tried to trade the captured territory for an Indian recognition of Chinese ownership of the territories they held before the border war began.

One might surmise that Chinese intentions with respect to their invasion of Vietnam in 1979 were similar.[17] The Chinese, knowing that Vietnam's recent alliance with the Soviet Union severely limited their freedom of action, had to produce whatever effect they could on Vietnamese policy quickly, before the Soviet Union could bring its superior military power to bear on the situation. This can be illustrated by Zbigniew Brzezinski's account of Deng Xiaoping's explanation of Chinese policy, which he gave to President Carter during his visit to the United States, which immediately preceded the invasion:

> China . . . had concluded that it must disrupt Soviet strategic calculations and that "we consider it necessary to put a restraint on the

---

[16]It is also worth noting that the initial fighting coincided with the Cuban missile crisis, during which the Soviet Union apparently felt constrained to support the Chinese. By November, the former Soviet Union had moved back to its previous position of pro-Indian "neutrality."

[17]It is of interest that the invasion of Vietnam was China's first major use of military force following the death of Mao Zedong.

wild ambitions of the Vietnamese and to give them an appropriate limited lesson."[18]

> He [Deng] then calmly diagnosed for [President Carter and his advisors] various possible Soviet responses, indicating how China would counter them. He included among the options "the worst possibility," adding that even in such a case China would hold out. All he asked for was "moral support" in the international field from the United States.

> China would undertake a limited action and then withdraw its troops quickly. Citing the Chinese-Indian clash of 1962 as an example, Deng insisted that the Vietnamese must be similarly punished. (Brzezinski, 1983, pp. 409–410.)

Deng, of course, understood the risk of Soviet intervention and sought to minimize it by emphasizing just how limited the military operation would be:

> We estimate that the Soviet Union will not take too big an action. . . .
> I think our action is limited, and it will not give rise to a very big event. (Cowan, 1979.)

Presumably, the Chinese hoped to achieve a major political effect by quickly defeating the Vietnamese forces and causing a certain amount of panic in Hanoi. In such an environment, the Vietnamese might have felt compelled to withdraw forces from Cambodia to bolster their defenses against the invading Chinese.[19] At best, Vietnam might have been led to rethink its policy of alliance with the Soviet Union and to retreat to a more balanced policy with respect to its relations with the Soviet Union and China.

Whether or not these goals could have been achieved had the operation gone well militarily for the Chinese is hard to determine. The basic Chinese dilemma—that its ability to increase the pressure on Vietnam was severely restricted by the possibility that the Soviets

---

[18]The Chinese may have also hoped that the invasion would "force Vietnam to withdraw some of its units from Cambodia." (Chanda, 1986, p. 358.)

[19]Some PRC sources have claimed that, in any case, the invasion prevented Vietnam from "devoting all its resources to eliminating the Pol Pot guerrillas." (Ross, 1988, p. 226.)

could take military action against China—would have remained in any case. If Vietnam retained a belief that Soviet involvement would save it before Chinese pressure became intolerable, it might have been able psychologically to withstand the type of shock that India experienced on its defeat in 1962. Of course, shaking Vietnamese faith in the Soviet Union was one of China's goals, but it was undercut by statements—such as Deng's claim that the limited nature of China's actions would preclude a Soviet response, cited above—that, if anything, emphasized the effectiveness of the Soviet alliance for Vietnam.

In the event, China's military operations were much less successful than those against India.[20] Despite having achieved initial tactical surprise, the Chinese advance soon bogged down, and Vietnamese defenders were able to hold off the initial Chinese attacks. The Chinese attacked along five lines of advance, each aimed at a provincial capital. The fighting reached a climax with the Chinese attack on February 27 on the provincial capital of Lang Son, which was the last major obstacle on the most direct route between China and the Red River delta (where Hanoi and Haiphong are located). The battle for Lang Son was difficult, involving house-to-house fighting. The Chinese finally succeeded in capturing the city and its surrounding hills on March 5; at that point, having demonstrated their ability to advance into the Red River delta and threaten Hanoi, they began to withdraw from Vietnamese territory.[21]

The Chinese were thus able to claim victory, but the difficulties and casualties they suffered in the process undercut the effectiveness of

---

[20]Vietnam, which had been fighting against Western armies for the better part of 30 years, was undoubtedly a much more serious military opponent than India.

[21]See Jencks (1979), pp. 801–815, for an account of the combat. The effects of this demonstration were, however, devalued by the fact that at least five of Vietnam's best divisions had not been committed but had remained in reserve for the defense of Hanoi. (On the other hand, China may have kept *its* best forces on the Soviet border.) In addition, China might have encountered serious difficulties in moving beyond the air defense umbrella that Chinese surface-to-air missiles based in China provided; in the open country of the Red River delta, Vietnamese fighter-bombers, longer-range tank guns, and anti-tank missiles could have taken a large toll on the Chinese troops. In short, in Jencks's assessment, "the PLA . . . had the opportunity to 'conduct a modern war' after the fall of Lang Son, but wisely declined it." (Jencks, 1979, p. 814.)

the "lesson" they were trying to convey.[22] Several months later, the Chinese deputy prime minister told American journalists that the invasion "did not give the Vietnamese enough of a lesson." (Li Xiannian, 1979.)

Opportunistic Timing

The timing of various uses of force depends heavily on the circumstances. In the cases discussed above, the Chinese were reacting to external events, which by and large dictated the timing of their own moves. In other cases, however, when the Chinese have taken the initiative in pursuing long-standing territorial claims, one can see a clear pattern of opportunistic timing. The key factor in timing appears to be the isolation of one's intended target from allies and other potential sources of support. In the cases discussed in this section, that isolation was a product of circumstances of the moment; one could also imagine cases in which a key part of the preparation for the attack would be an attempt to create difficulties between the intended target and its allies.

On three occasions during the past 25 years, China has used force to take possession of islands in the South China Sea:

- the Crescent group in the Paracel (Xisha) Islands in 1974 (from South Vietnam)

- Johnson Reef in the Spratly (Nansha) Islands in 1988 (from Vietnam)

- Mischief Reef in 1995 (from the Philippines).[23]

In each of these cases, the timing was opportune. The South Vietnamese government had been essentially abandoned by the United States by 1974 (and, in any case, the United States did not want to jeopardize its *rapprochement* with China over such a minor

---

[22]The Chinese admitted to 20,000 casualties, including 10,000 dead; Vietnam claimed Chinese losses of twice that amount. (Segal, 1985, p. 219.)

[23]Unlike the other two cases, this situation did not result in an actual military conflict between the two sides; the Chinese stationed armed vessels at and built structures on Mischief Reef without being observed. The Chinese did detain Filipino fishermen in the region.

issue). North Vietnam, on the other hand, could hardly complain about an attack on the forces of its enemy, although it must have been clear that China's action would redound to the ultimate disadvantage of the unified Vietnam the North Vietnamese leadership was in the process of creating.

By 1988, Soviet support for Vietnam was faltering, both because of changes within the Soviet Union itself and because of Gorbachev's hopes for some sort of reconciliation with China. At the same time, Vietnam was still far from having mended its fences with the West, which would only occur after it had withdrawn from Cambodia and reduced its support for the Hun Sen government it had installed in the wake of its 1978–1979 invasion. Similarly, the Philippine government in 1995 was more isolated than it had been previously because of the cancellation of the U.S. leases on military bases at Clark Field and Subic Bay.

CHARACTERISTICS OF CHINESE USE OF FORCE IN A CRISIS

In addition to using force in actual conflicts, the PRC has used force during a crisis several times[24]; indeed, one could say that China typically used force *to create* a crisis.[25] It is often claimed that, since *weiji*, the Chinese word for *crisis*, is composed of two characters that can be translated as *danger* and *opportunity*, respectively, it does not have an entirely negative connotation. This is apparently apocryphal[26]; nevertheless, it does appear that the Chinese leadership

---

[24]The dividing line between "crises" and actual conflicts is admittedly a vague one; most of the crises discussed below involve the use of force in a way that goes beyond a mere exercise or show of force. However, they are discussed under the heading of "use of force in a crisis" because of the judgment that the political purposes of the action did not depend on the achievement of any military goal.

[25]As used here, *crisis* has a military connotation and is defined by two characteristics: the use of the "military element" of national power in some fashion (alerts, unusual deployments, or actual firing of weapons) and the belief that the there is an increased probability of the outbreak of armed conflict. Thus, the *initiation* of the crisis means the first use of the military element; for example, while the PRC may have considered that Taiwanese president Lee Teng-hui's visit to the United States in 1995 created a critical political situation, the actual (military) crisis did not begin until the PRC initiated military exercises designed to intimidate Taiwan.

[26]Or so the authors understand from a native speaker.

does not necessarily consider a crisis to be a negative phenomenon: It may provide an opportunity for making gains that would otherwise not be achievable.[27]

One common reason for wanting to create a crisis is to consolidate domestic political standing and to mobilize the population in support of the regime's goals. Thus, one interpretation of the 1958 Taiwan Strait crisis is that it served to help mobilize the Chinese people for the exertions of the Great Leap Forward, which was just beginning at that time.[28] According to this view, Mao Zedong wanted to instill into the population, for the purpose of implementing the Great Leap Forward, the same spirit of sacrifice and patriotism that had animated the struggle against the Japanese. As Mao explained to a party conference on August 17, shortly before the initial shelling began on August 23,

> In our propaganda, we say that we oppose tension and strive for détente, as if détente is to our advantage [and] tension is to their advantage. [But] can we or can't we look at [the situation] the other way round: is tension to our comparative advantage [and] to the West's disadvantage? Tension is the West's advantage only in that they can increase military production, and it's to our advantage in that it will mobilize all [our] positive forces. . . . Tensions . . . can [help] us increase steel as well as grain [production].[29]

Aside from domestic political reasons, the Chinese have initiated crises to probe the reactions of adversaries[30] and to impress on them the costs—political and potentially military—of pursuing a policy hostile to China's interests. A crisis brings home to the adversary government—and, in particular, its population—the possibility that its policies could lead to war; it may also increase tensions and

---

[27]The view that crises are invariably bad—that they present above all "danger" rather than "opportunity"—is the understandable perspective of a *status quo* power, such as the United States, which seeks primarily to avoid war, especially nuclear war.

[28]For a compelling argument in favor of this interpretation, see Christensen (1996b), Ch. 6.

[29]Mao Zedong speech to the Beidaihe Party Conference, August 17, 1958, in MacFarquhar et al. (1989), p. 402.

[30]In an earthy aphorism attributed to Mao, "You can't know the reaction of the tiger if you don't touch his ass." (Segal, 1985, p. 211.)

disagreements among the members of the hostile alliance or coalition.

The first Taiwan Strait crisis of 1954–1955 falls into the latter category. On July 23, 1954, as Premier Zhou Enlai was returning from the Geneva conference on Indochina, Chairman Mao Zedong sent him the following message:

> In order to *break up* the collaboration between the United States and Chiang Kai-shek, and keep them from joining together militarily and politically, we must announce to our country and to the world the slogan of liberating Taiwan. It was improper of us not to raise this slogan in a timely manner after the cease-fire in Korea. If we were to continue dragging our heels now, we would be making a serious political mistake. (Zhang, 1992, p. 193; emphasis added.)[31]

In essence, China attributed to the United States a "three front" strategy; the United States sought to exert military pressure on China from Korea, Vietnam, and Taiwan. By mid-1954, the first two "fronts" had been dealt with. What remained was to try to do something about the potential third "front" (Zhang, 1992, p. 189): China may have feared that a formal U.S.-Taiwan alliance was an obvious next step following the formation of the Southeast Asia Treaty Organization and the signing of a mutual defense treaty with South Korea.[32]

The shelling of Jinmen (Quemoy) and Mazu (Matsu) served to create political pressure on the Eisenhower administration, which was presented with the following dilemma: either disinterest itself in the fate of the offshore islands, thereby potentially undercutting the prestige and morale of the Nationalist Chinese regime on Taiwan, or follow a policy that appeared to risk a major war on account of some islands that, from the point of view of American public opinion, were far away and generally insignificant. The latter choice would create

---

[31]Zhang cites an unpublished paper by He Di, a member of the Institute of American Studies, Chinese Academy of Social Sciences. See also He Di (1990), pp. 222–245.

[32]This is the interpretation put forward by He Di (1990), pp. 224–225:

> With the signing both of a mutual defence treaty with South Korea and the protocol creating the Southeast Asia Treaty Organization the U.S. government entered into negotiations with the Kuomintang to form a mutual defense treaty—the last link in the ring of encirclement of China.

domestic political difficulties; it also ran into the opposition of U.S. allies in Western Europe, who feared that, if the United States were to become bogged down in a war against China, it would be forced to neglect the Europeans' security concerns vis-à-vis the Soviet Union.

If Mao hoped to break up the incipient U.S.-Taiwanese alliance by means of the crisis, his action backfired. In fact, both President Dwight D. Eisenhower and Secretary of State John Foster Dulles had been reluctant to conclude a mutual security treaty with Taiwan. Eisenhower thought that such a treaty would be "too big a commitment of U.S. prestige and forces."[33] Nevertheless, under the pressures of the crisis, U.S. policy shifted toward a treaty with Taiwan, which was concluded on December 5, 1954.

A related goal for the sake of which one might initiate a crisis would be to throw one's adversary off balance, to "attack the enemy's strategy."[34] This may have been part of the motivation for the Chinese ambush of Soviet border troops on Damansky-Zhenbao island on March 2, 1969. Hypothetically, the Chinese may have believed that the Soviet policy was to increase military pressure on the border gradually until the Chinese were willing to cave in to Soviet political demands.[35] In such a situation, the Chinese may have decided that a dramatic event was necessary to make clear to the Soviets that China

---

[33]As recorded in Zhang (1992), p. 204, who cites as his source the Dulles papers, Conference with the President, May 23, 1954, White House memorandum series, Box 1, Eisenhower Library.

[34]In accordance with the well-known advice of Sun Zi (3:4–7):

Thus, what is of supreme importance in war is to attack the enemy's strategy.

Next best is to disrupt his alliances.

The next best is to attack his army.

The worst policy is to attack cities. Attack cities only when there is no alternative.

[35]According to one assessment, a rough balance existed between Chinese and Soviet forces along the border until about 1964; after that, and especially after 1966, when Soviet troops were stationed in Mongolia, the balance shifted to the Soviets' favor:

by early 1969 . . . the magnitude and continuation of the Soviet buildup more than offset [China's gain from the end of the disruptions of the Cultural Revolution] and probably caused the Chinese to be fearful of the future. (Robinson, 1970, p. 32.)

was willing to risk war rather than submit to a gradual loss of political maneuvering room.[36]

A generally weaker power using force in this way must calibrate it very carefully. The force used, plus the accompanying rhetoric, must be sufficient to create an acute sense of crisis, since the political effect depends on it. On the other hand, it must not be so great as to prompt a stronger reaction than the initiator intended; obviously, there is a danger that unintended conflict will result.

---

[36]This is a variant of one of the explanations Thomas Robinson proposed, which understood the ambush as a Chinese effort to "preempt" the Soviets, i.e., to head off a clash that would otherwise become inevitable. But if the Soviets were willing to fight a war with China, all China was doing was providing an opportunity and excuse; if, however, the Soviets hoped to get their way without war, merely by increasing the military pressure on China, the March 2 incident would have forced them to rethink their strategy. See Robinson (1970), pp. 53–55, for possible explanations of Chinese behavior.

Chapter Three

CHINESE NATIONAL MILITARY STRATEGY

Having briefly reviewed the PRC's historical use of force as a background, we turn to Chinese doctrinal writings on these questions for more insight into how the Chinese might view any future use of force on their part, especially either against the United States or in circumstances under which U.S involvement in the resulting conflict could not be ruled out. We look first at the history of Chinese national military strategy, as articulated in Chinese writings; as already noted, the evolution of national military strategy reinforces the impressions gained from the historical review concerning how the Chinese might use force.[1]

WHAT IS NATIONAL MILITARY STRATEGY?

By *national military strategy*, we mean a nation's authoritative view of the most effective way to organize and employ the nation's military forces to protect or pursue its interests in the existing international environment.[2] This strategy depends decisively on a view about the nature of the most important type of future war in which the nation is likely to be involved and guides decisions concerning the nation's preparations for war (in terms of weapon procurement, the development of combat methods, training, etc.). The former

---

[1]The discussion of PRC *national* military strategy applies, of course, to the post-1949 period. However, as noted later in the text, the key concept of "people's war" reflected the Chinese Communist Party's (CCP's) pre-1949 experiences fighting against the Guomindang (GMD) regime and the Japanese.

[2]According to then–CMC Vice-Chairman Liu Huaqing, a nation's military strategy is "the basic grounds for building and using armed forces." (Liu Huaqing, 1993, p. 15.)

Soviet military used the term *doctrine* to describe such a set of views about future war, and this term is often used to refer to the Chinese view of future war as well; we have not done so here because the term *doctrine* is used differently in U.S. military terminology.

It is important to note that national military strategy, in this sense, deals with views concerning *future* war. Of course, the war or wars that a nation actually fights in a given period may be quite different from the type of future war that most prominently guided its military preparations. Thus, the United States prepared during the Cold War for a major conventional and potentially nuclear war with the Soviet Union but in fact fought wars of a different sort in Korea and Vietnam. Similarly, in the first three decades of its existence, the PRC prepared for major invasions by the United States and, later, the Soviet Union, but actually fought very different types of wars, first, in Korea (1950–1953) and, later, against India (1962) and Vietnam (1979).

EVOLUTION OF THE PRC'S NATIONAL MILITARY STRATEGY

From 1949 to the late 1970s, China's national military strategy was defined by Mao Zedong's concept of "people's war." However, like many other aspects of Chinese state and society, this changed dramatically following Mao's death. With the onset of the Deng Xiaoping era, Chinese strategists began adjusting the PLA's strategic and operational doctrines according to perceived changes in the nature of the threats facing China and the military capabilities required to meet them. By the early 1990s, China's national military strategy shifted to preparing to engage and win a "local war under high-tech conditions." This type of war is dramatically different from the type of warfare with which Mao was dealing when he developed the "people's war" concept.

People's War

The essence of "people's war" has been described as "a military strategy designed to turn weaknesses into strength by utilizing China's vast size and manpower to wage a war of attrition against an invader." (Garver, 1993, p. 257.) Mao Zedong developed the concept of "people's war" during the Chinese Civil War (1927–1937, 1945–

1949) and the Anti-Japanese War (1937–1945). The "people's war" strategy was based on the perception that China faced a near-term threat of invasion by a materially superior, but numerically inferior, adversary that sought to occupy substantial areas of Chinese territory. Throughout much of the Cold War, "people's war" was held to be the most effective means of addressing what the Chinese believed to be the threat of invasion by one of the two superpowers (the United States in the 1950s and early 1960s and the former Soviet Union from the late 1960s to 1985).

The central aim of "people's war" was to drive out or annihilate invading enemy forces by means of a protracted war of attrition, in which regular and irregular (guerrilla) Chinese forces would eventually exhaust and overwhelm the numerically inferior, overextended invading armies. In this way, China could blunt the enemy's advantages in technology and firepower and bring to bear China's own advantages of strategic depth and large population. As Mao noted regarding invading Japanese forces,

> everybody would be in favor of driving the "devils" out overnight. But we [the CCP] point out that, in the absence of certain definite conditions, quick victory is something that exists only in one's mind and not in objective reality. (Mao, 1975b, p. 133.)

The "objective reality" was that China was a "large and weak" country that could only hope to achieve victory against superior invading forces "through the cumulative effect of many offensive campaigns and battles in both regular and guerrilla warfare." (Mao, 1975b, p. 84.)

Mao outlined three phases of "people's war." The first stage encompassed the enemy's initial strategic offensive. (Mao, 1975b, p. 137.) During this stage, the adversary would be able to take advantage of its material superiority to score substantial victories. Rather than attempt to defend fixed positions against overwhelming odds, Chinese forces in this stage would rely on "mobile warfare," which may be defined by Mao's formula, "fight when you can win, move away when you can't." (Mao, 1975a, p. 241.) This allowed Chinese Communist forces to take advantage of situations in which they could achieve a local superiority over the adversary, while avoiding direct confrontations at times when the enemy could bring its superior firepower to bear.

Mao characterized guerrilla warfare as an embryonic form of mobile warfare (Mao, 1975a, p. 243), marked by the use of irregular troops and a decentralized organization. Guerrilla warfare is highly flexible and fluid. It is most suitable for a military force in the early stages of its development or for one that must operate in an area where the enemy holds a clear and decisive advantage. Mao expected guerrilla warfare to evolve into mobile warfare as the war progressed.

The second stage was marked by strategic stalemate and is the transition stage of "people's war." Following its early successes, the enemy will have to expend increasing amounts of resources to consolidate its gains. Fewer enemy troops will be available for offensive action as more and more are required for such duties as occupying cities and maintaining secure supply lines.

At this stage of the conflict, China would rely primarily on guerrilla warfare in occupied areas to hamper the enemy's consolidation efforts. As Mao explained in reference to Japan in the 1930s,

> scores of her divisions will be inextricably bogged down in China. Widespread guerrilla warfare and the people's anti-Japanese movement will wear down the big Japanese force, greatly reducing it and also disintegrating its morale by stimulating the growth of homesickness, war-weariness and even anti-war sentiment. (Mao, 1975b, pp. 138–139.)

The most important task during this stage, however, is to mobilize as much of the population as possible to sustain and carry forward the war against the enemy. Since Chinese forces would remain too weak to assume the strategic offensive, they must enlist the efforts of the great masses of Chinese people in the struggle. As Mao explained with regard to the Japanese invasion,

> the mobilization of the common people throughout the country will create a vast sea in which to drown the enemy, create the conditions that will make up for our inferiority in arms and other things, and create the prerequisites for overcoming every difficulty in the war. (Mao, 1975b, p. 154.)

The third stage encompasses the final strategic counteroffensive by Chinese forces to regain lost territories and drive enemy forces out of China or annihilate them altogether. Mobile warfare would again be the primary form of warfare, with positional warfare assuming increased importance and guerrilla warfare reduced to a supplemental role. The PLA finally entered the third stage of "people's war" in 1947 in their resumed conflict with Chiang Kai-Shek's GMD forces.[3]

Ultimate success in "people's war" depended on the defending side's ability to mobilize the population. Military professionalism and expertise were less of a priority, therefore, than the ability to mobilize politically the largest possible percentage of the population against the adversary. This reflects the importance of the great disparity in popular support between CCP and GMD forces during the renewed Civil War between 1945 and 1949 for the PLA's ultimate success. Although there was considerable debate within the CCP leadership following "liberation" in 1949 regarding the role of professionalism in the PLA, the emphasis on being "red" rather than "expert" was maintained, reaching a high point during the Cultural Revolution.[4]

The massive Third Front defense industrialization campaign of the 1960s and early 1970s reflected China's adherence to the "people's war" strategy.[5] The Third Front campaign was an attempt to establish an industrial base deep within China's hinterland that the PLA could rely on should it be driven from China's longer-established industrial centers in the east. These areas would function as a secure base area, similar to those established by communist forces in their long fight against the GMD and Japanese. However, unlike earlier base areas, Mao intended these to have the industrial capacity to support more than simple light-infantry forces. These preparations were most intense between 1964 and 1971, when tensions between China and both the Soviet Union and the United States were high.[6]

---

[3]Following World War II, Nationalist forces, with U.S. assistance, quickly reoccupied much of the country. GMD efforts in this regard appeared successful through 1945 and 1946, even driving the CCP out of its wartime stronghold of Yenan. However, by 1947, their forces were overextended, particularly in the northeast, and quickly succumbed to Communist counteroffensives.

[4]As early as the 1950s, a number of leaders within the PLA, Peng Dehuai most notably, advocated a more modernized, professional military based on the Soviet model. This was an important area of disagreement between Mao and Peng and contributed to Peng's downfall at the Lushan Conference in 1959. (Lieberthal, 1993, p. 101.)

[5]It is worth remembering, however, that, despite the emphasis on "people's war," China successfully pursued a strategic nuclear capability during this period.

[6]For greater detail on the Third Front campaign, see Naughton (1998).

People's War Under Modern Conditions

China began to shift formally from "people's war" to the newer "people's war under modern conditions" strategy in the years immediately following Mao's death.[7] In 1978, China's military leadership determined that its traditional military strategy was no longer sufficient to deal with the contemporary Soviet military threat. (Jencks, 1984, p. 308.) Then–Defense Minister Xu Xiangqian highlighted the need for change in a 1979 article appearing in the ideological journal *Hongqi*:

> War is now conducted in a way different from that in the past. . . . The target of attack, the scale of war and even the method of fighting are new to us. . . . Our military thinking must tally with the changing conditions. If we treat and command a modern war in the way we commanded a war during the 1930's and 1940's, we are bound to meet with a big rebuff and suffer a serious defeat. (Xu, 1979.)

More specifically, "people's war under modern conditions" was intended to address the security challenges posed by the possibility of a Soviet attack across China's northern border. This is, ostensibly, the exact kind of threat "people's war" was designed to address. However, steady improvements in military technology, which by the late 1970s granted Soviet forces unprecedented accuracy, range, and destructiveness of firepower, brought the efficacy of the PLA's traditional "people's war" strategy into serious question. Chinese strategists observed that the Soviet Union had the capability to launch an

---

[7]Indeed, Mao's death was probably necessary for the PLA to move away from his "people's war" concept. Many of the concepts encompassed by "people's war under modern conditions" had existed within Chinese strategic circles for years. The destructive potential of Soviet forces, both conventional and nuclear, was evident well before 1978, as was the importance and vulnerability of China's industrial centers in the northeast. However, China's ideological climate from 1966 to the time of Mao's death rendered it very difficult for even the PLA to appear to abandon a concept so closely identified with Mao. As Joffe (1987) explains, "Tampering with Mao's military doctrine was a highly sensitive matter because it was inextricably linked to the broader issue of how to treat Mao's ideological legacy." The need for "circumspection" in discussing strategic military matters persisted until Deng Xiaoping's famous "seek truth from facts" speech of June 1978. Deng's speech allowed the PLA, as well as other institutions within China, to move beyond Maoist concepts that were no longer useful without appearing to repudiate Mao himself.

attack with little warning and harness a decisive advantage in fire-power during the initial stages of the conflict. Mechanization and the improved ranges and accuracy of modern weaponry enabled Soviet forces to concentrate rapidly and strike China simultaneously from the land, air, and sea. Finally, modern weaponry, nuclear weaponry in particular, held unprecedented destructive potential that could be directed at China's military, economic, and political centers with catastrophic effect. (Jencks, 1984, p. 310.)

"People's war under modern conditions" was also intended to pro-vide the PLA with an alternative method for dealing with future security threats that fell short of a general invasion. Even for the Soviet Union, the costs associated with attempting to invade and occupy all of China were prohibitive. However, it was feasible for Soviet forces to undertake a limited invasion of China and possibly occupy a part of China lying along the Sino-Soviet border, such as the industrially important northeast region. The former Soviet Union might also undertake limited military raids into Chinese terri-tory to punish or pressure Beijing. (Jencks, 1984, p. 307.) Mao's con-cept of a "people's war" did not provide China's military with viable options for dealing with these kinds of scenarios.

At the same time that Chinese leaders noted the disturbing implica-tions modern military technology held for their ability to defend against a Soviet attack, they also had to take into account the impor-tant changes that had occurred in China's own economy. By the 1970s, China had developed important economic and industrial centers in its northeastern and eastern regions. Not only did these centers represent a large part of China's productive capacity; they also supported the production of increasing numbers of modern weapon systems. Since the 1960s, China had invested heavily in developing its own production capabilities for relatively modern military systems, such as tanks, artillery, and aircraft. While most of these weapons were based on 1950s-era Soviet technology, they were, nonetheless, increasingly prominent components of the PLA force structure. The implication of these two developments was that the loss of industrial centers to an enemy invasion would be a far more severe blow to China's ability to sustain a war effort than it had been during the War of Resistance against Japan. (Garver, 1993, p. 259.) This remained true despite the Third Front campaign, which was intended to create self-sufficient industrial bases in the more-secure interior regions of China.[8]

"People's war under modern conditions" attempted to address the advances in the Soviet Union's, or other potential adversary's, offensive capabilities in a manner that did not require the automatic abandonment of important economic and political centers. It placed much greater emphasis on positional warfare, combined arms tactics, and the use of regular and mechanized forces to blunt an enemy's invasion before it could penetrate deep into Chinese territory and implied a much greater emphasis on logistic support. It sought to meet the new demands of "modern" warfare, conventional or nuclear warfare "conducted by using modern weapons and modern technology."[9]

A number of distinct characteristics of modern war stem from improvements in the mobility, range, and destructiveness of modern weapon systems. With respect to some characteristics, the notion of modern war merely represents a catching up with the realities of mechanized warfare as it existed in World War II. Thus, according to one Chinese analyst, most battles in previous wars were fought on the ground, with air and naval forces relegated to support duties. (Zong, 1983.) In earlier conflicts, battles were generally initiated in border regions, but in modern warfare, conflict can begin simultaneously in front and rear areas. Modern air and naval forces, in particular, are able to launch strategic assaults into an adversary's rear, making it more difficult for the attacked country to maintain production and mobilize human and material resources for the conflict.

As a result, the initial stage of conflict gains in importance. By concentrating firepower more accurately over a shorter period of time with a greater destructive effect than in the past, an aggressor is able to inflict much more severe damage on defending forces than was possible in past wars, particularly if it uses nuclear weapons, and is thereby able to achieve a decisive military advantage in the early

---

[8]For all the resources poured into the Third Front campaign, China's main industrial centers remained in the east and northeast regions. Again, see Naughton (1998).

[9]For a Chinese analyst's discussion of the distinctive features of modern war, see Zong (1983).

stages of conflict. As a result, "the battles in the initial stages will play a more important role than before in winning the initiative of the war and its later development." (Zong, 1983.)

Finally, modern war consumes larger amounts of war materials and equipment than did traditional "people's war," which implies an increased dependence on logistics. It is no longer feasible for China to defeat an invading enemy with an army of "rifles plus millet," to quote Mao's description of the early PLA. To fight and win a modern war, therefore, the PLA requires modern logistical support and therefore must be able to mount some defense of important industrial areas. (Zong, 1983, pp. 78–83.)

China's development of a strategic nuclear capability in the early 1970s was also an important factor in China's evolving strategic calculus.[10] Despite being only a fraction of the size of the two superpowers' nuclear forces, China's nuclear arsenal represented a powerful deterrent. Its existence rendered a full-scale invasion of China prohibitively costly to any would-be aggressor.

Even with the changes in China's military strategy described above, however, important elements of the "people's war" strategy remained. The Soviet Union remained China's primary security threat. Chinese forces were still unlikely to be able to stop the better-equipped Soviet army at the border. Victory over a large-scale Soviet invasion would, therefore, still require a prolonged and costly campaign of attrition. (Jencks, 1984, p. 315.)

THE SHIFT TO "LOCAL WAR UNDER HIGH-TECH CONDITIONS"

In late spring 1985, China's CMC instructed the PLA that it was no longer necessary to prepare for "early war, major war, and nuclear war" with the Soviet Union. Instead, the CMC declared that PLA military strategy should focus on preparing to fight and win a "local" or "limited" war (*jubu zhanzheng*). Following the Gulf War, Chinese strategists added the phrase "under high-tech conditions" to their

---

[10]Although China's first nuclear test occurred in 1964, its first ballistic missile, the DF-3 intermediate-range ballistic missile, entered service in 1971. (Khalilzad et al., 1999, p. 43.)

doctrinal terminology. Thus, by the early 1990s, the PLA was charged with preparing to fight a "local (or limited) war under high-tech conditions."

The changes in China's military strategy in the mid-1980s and early 1990s resulted from the perception of mutually reinforcing changes in the nature of security threats confronting the PRC and the role of technology in military conflict; there was a reassessment both at the strategic level (the type of war to be fought) and at the operational level (how a war would be fought). In 1985, the CMC declared that China no longer faced a serious invasion threat from either of the two superpowers. (Li, Nan, 1996, p. 445.) It believed that, due to the military buildup the Reagan administration had undertaken, the United States had a slight edge in its strategic competition with the Soviet Union. However, given the nuclear stockpiles in both countries, Chinese strategists concluded that the superpower rivalry had reached a strategic stalemate, a point at which neither the United States nor the former Soviet Union could risk attacking the other or, more importantly, China. (Godwin, 1992, p. 192.) As Li Desheng, former commandant of China's National Defense University, explained,

> It can be said that the Soviet Union and the United States, which are capable of fighting a major war, have achieved parity of strength, with neither enjoying superiority over the other. Neither of them would dare take reckless action.[11]

Moreover, Chinese analysts argued that the two superpowers were economically impoverishing themselves through their bipolar duel. This, taken in conjunction with the rise of important regional powers, Japan and Germany in particular, was hastening the emergence of a multipolar world and an overall decline in the two superpowers' influence on world affairs.

The absence of an impending threat of invasion did not, however, translate into a totally benign security environment for the PRC. One result of diminishing Soviet and U.S. influence over world events was that regional tensions, long held in check by the Cold War super-

---

[11]"Shift in China's Strategy for Building Armed Forces Shows New Evaluation of World Situation" (1986), p. 1.

power rivalry, once again were able to assert themselves. Chinese analysts noted that, in China's immediate region, many conflicts existed related to political disputes, disputes over maritime territory, and disputes between countries with different social systems. The decline of superpower influence over the regional actors involved in these disputes, they argued, "will lead to an increase of partial and regional military conflicts and medium- and small-scale wars." (Jia, Zheng, and Guo, 1987.)

Chinese strategists identify five types of "local war": (1) small-scale border conflicts, (2) contention for territorial seas and islands, (3) surprise air attacks, (4) resistance against partial hostile intrusions, and (5) punitive counterattacks. (Jia, Zheng, and Guo, 1987.) This list appears to be based on recent experience, Chinese and foreign, and it is possible for a given limited war to demonstrate features from more than one of the "types" listed above. Common to all "types" of local war, however,

> are the limited political objective ... and the requirement ... to respond quickly either to defeat the presumed political purpose of the attack or to gain the political objective sought by the limited use of force. (Godwin, 1992, p. 194.)

Past PRC actions, such as its intervention in Korea in 1950, illustrate China's willingness to use large-scale force and accept high casualties in pursuit of its interests. Nevertheless, the current Chinese military strategy of "local war under high-tech conditions" evinces a belief that current international conditions place substantial constraints on the use of force: A "local war" is a conflict circumscribed in both the geographic scale in which it occurs and the scope of the political objectives pursued. (Jiao and Xiao, 1987.) Unlike Mao's concept of "people's war," the purpose of military force in local warfare is not to annihilate or drive out the enemy. Rather, it is

> to assert one's own standpoint and will through limited military action, or to teach the opponent a lesson, to deliver a shock, politically or psychologically, and then immediately withdraw to home bases. (Jiao and Xiao, 1987.)

The use of military force under local war conditions, therefore, serves to punctuate or amplify the nation's diplomacy. Political and diplomatic forces play a far greater role in determining the course and outcome of limited war.

The potential for an escalation or expansion of such a conflict is constrained by such factors as the enormous costs of conducting the war, the impact of critical public and world opinion, and the threat of nuclear weapons.[12]   Just as the threat to China is considerably reduced in the post–Cold War era, so too is the latitude that states (including China) have for using force in pursuit of their interests. The nature and consequences of these constraints, to which we now turn, help explain the notion of local war.

Primacy of Economic Development

According to many Chinese observers, a key feature of the current international system, especially with respect to East Asia, is the primacy that most states now place on economic development. To some extent, this merely reflects the end of the Cold War, which was characterized by a militarized political-ideological rivalry between the two blocs.[13]   As a result, economic interests have come to the fore; while this does not rule out all conflict and confrontation, it places it in a different context because of the increasing globalization of the world's economic system. Thus, while nations will still compete for political and economic advantage, they will do so in the context of an international economic system that is becoming increasingly interdependent and integrated. This is certainly the case for China itself, which depends heavily on foreign markets and sources of investment capital and, increasingly, on imported sources of energy as well. Until adequate pipelines and other transportation facilities can be built to carry oil overland from Central Asia, this involves a dependence on the seaborne transport of oil.

Thus, for China, economic development is now seen to be dependent on the opening to the rest of the world set in place by Deng Xiaoping's policies. This acts as a constraint on the use of force; to

---

[12]Further discussion of these factors can be found in the following: Shi Yukun (1995) and "Shift in China's Strategy for Building Armed Forces Shows New Evaluation of World Situation" (1986).

[13]This point is made in Zhao Ying (1996).

the extent that conflict would disrupt the relationships on which economic development depends, it would carry a high price for the Chinese leadership.

Similar considerations apply to neighboring countries as well. The economic development of the entire East Asian region has been based on increased foreign trade and investment; any conflict in the region could have serious negative consequences for all the nations in the area. This would increase pressures to end any conflict quickly.

Indeed, even potential adversaries are likely to have important economic links that will create pressures for the rapid cessation of any hostilities. Taiwan has already become a major investor in China; this relationship survived the tensions surrounding the Chinese military exercises in early 1996. Thus, any future conflict between China and Taiwan will take place in the context of cross-cutting pressures, created by the trade and investment links, to de-escalate the conflict and find some resolution to it before its economic costs become excessive.[14]

Nuclear Weapons

Another factor leading to the damping of conflict has been the existence of nuclear weapons and hence the risk that even small conflicts between nuclear powers could, under some circumstances, escalate to the nuclear level. China's confidence, since the mid-1980s, that it does not face a major threat to its survival presumably reflects not only the political changes in the Soviet Union (the coming to power

---

[14]Historically minded readers will perhaps recognize echoes of Norman Angell's *The Great Illusion* (Angell, 1912), which argued that the economic interdependence of modern economies made war among the European powers unprofitable and hence irrational; first published in 1909 under the title *Europe's Optical Illusion*, this book has become a byword for misplaced optimism. However, the argument here is somewhat different in two important respects: First, the primacy of economic development is regarded as something that is recognized by the governments involved themselves, and, second, as we shall see, the argument is not that armed conflict is impossible or even unlikely but merely that tremendous pressures will be exerted on the participants to end it quickly. Of course, this is no guarantee that these pressures will in fact cause an early termination of a conflict: Wars often last much longer than their initiators thought likely, and combatants often continue to fight long after economic rationality might have suggested that they stop. World War I itself is a case in point.

of Gorbachev) but also its acquisition of a reasonably secure second-strike strategic nuclear capability.[15] Even conflicts involving a nonnuclear power are potentially affected by this dynamic, if the nonnuclear power has a nuclear-armed ally.

Other Factors

In addition to these factors, some Chinese observers also talk more generally about the "trend of the times" and "global opinion" as constraints on the international behavior of states. This is similar to what in the West is called a "liberal" (as opposed to a "realist") theory of international relations.[16] However, "liberal" theories tend to emphasize the importance of multinational organizations as sources of restraint on state sovereignty; such organizations appear to play a lesser role in the Chinese understanding, which more typically sees an encompassing web of bilateral relationships and interdependencies. As a *People's Daily* editorial expressed it,

> a relationship of interdependence, mutual support, and mutual containment is developing between big nations. This dictates that big nations be restricted by multilateral relations when handling bilateral relations. This mutual check and balance is conducive to the relative stability of the relations among big nations. (Zhang and Zhu, 1997.)

This is admittedly a vague point, and it is hard to tell whether it should be accorded any independent weight. Nevertheless, it seems worth including as a perhaps subsidiary element of the overall worldview underlying the "limited war" concept.

---

[15]Indeed, the shift in the late 1970s to the "people's war under modern conditions" concept may already have reflected China's confidence that the Soviet Union would not be willing to risk an all-out invasion of China but would set itself a smaller objective, one presumably below China's "nuclear threshold."

[16]A particularly striking exposition of this "nonrealist" view argued not only that "[i]n today's world, exchange and cooperation have become the general trend" but went on to argue that a new "international regime" is emerging, comprising

> many good international systems [international trade regulations, Law of the Sea, etc.] which are followed by most countries [and which make] it possible to reduce the number of conflicts and bring about a better order in the world. (Tang, 1996.)

Net Effect of These Constraints

The net effect of this somewhat nebulous collection of factors is to constrain the actions of nations in the international arena—even of the most powerful nation, the United States—and to force states involved in conflicts to seek as rapid a resolution of them as possible. It might also seem likely that these forces would tend to induce states not to resort to the use of force in the first place.[17] While there may be some truth to this, the standard argument often overlooks the fact that the same restraints that operate on the initiator of a given use of force also (more or less) operate on the country against which the force is being used.

Thus, while China might have to consider the world reaction to potential uses of military force to intimidate Taiwan, Taiwan and the United States would have to take the same sorts of considerations into account in considering how to respond to them. Thus, both sides are similarly constrained in their use of force, and both—other things being equal—have a similar incentive to bring the crisis to a close as soon as possible. Consequently, while the initiator may feel constrained in his use of force, he may also be encouraged by the belief that his target (or its allies) will be likewise constrained in their response.

To some extent, the situation resembles a version of the game of "chicken" as analyzed by Herman Kahn: The side that is willing to run the greater risk (in this case, proceed further in courting the risk of obstructing economic relationships and otherwise disturbing the relationships of interdependence) has an advantage. The key lesson, however, is that it will be necessary to achieve whatever goal one is pursuing quickly; if one can do that, one is in the advantageous position of presenting one's opponent with a *fait accompli*, thus transferring to him the onus of prolonging the conflict.

---

[17]This indeed is the "liberal" international theory argument about the effects of the growth of international institutions.

Chapter Four

CHINESE USE OF FORCE IN THE FUTURE

In thinking about possible future PRC uses of force, we must take into account that China is likely to gain in military strength during the next several decades.[1] Nevertheless, it will remain considerably weaker than the United States in overall military terms. Thus, in terms of what may be the most important planning scenario for the PRC, i.e., military action against Taiwan in the face of U.S. opposition, China will have to continue to focus on the problem of confronting a stronger power.

EXTRAPOLATING TO ACCOUNT FOR A STRONGER CHINA

The historical record of PRC use of force typically involves the use of force against a stronger power or against the client of a stronger power. The exceptions are the Sino-Indian border war of 1962 and the instances of Chinese use of force in the South China Sea against Vietnam and the construction of facilities on Philippines-claimed Mischief Reef. Similarly, Chinese strategic writings often concentrate on the situation of a weaker, less technologically advanced China confronting a stronger, more technologically advanced adversary.

In using force against the United States, the Soviet Union, or their clients, the Chinese had to be mindful of the possibility that their actions, although meant to be carefully calibrated, could have been seen as more threatening than intended. This, of course, ran the risk that China could find itself involved in a major war against a much

---

[1]See Khalilzad et al. (1999), Chapter Three, for a discussion of the ways in which China's military strength can be expected to increase.

more powerful adversary. The PRC has adopted various strategies to deal with this problem.

During the period of its close alliance with the Soviet Union, China sought to rely on the deterrent effect of Soviet strategic nuclear forces to protect it against an all-out attack by the United States. In the aftermath of its break with the Soviet Union, the PRC complained that it had received insufficient support in this respect.[2] In the 1960s, the PRC was without a superpower ally, and, even after the *rapprochement* with the United States in 1972, the degree to which it could rely on U.S. strategic nuclear forces to deter the Soviet Union was uncertain. "People's war" played an even more important role. Since this concept claimed to describe a way for a weaker China to be victorious eventually, the prospect of major war with a more powerful adversary could be faced with confidence.[3]

The Chinese may now feel that their own strategic nuclear forces can now provide a sufficient deterrent against large-scale attack. It is true that the Chinese "no-first-use" pledge implies that its nuclear forces are useful for deterring only a nuclear attack. However, the Chinese may believe that an adversary, despite its greater conventional strength, would nevertheless regard a large-scale conventional attack as infeasible. Alternatively, the Chinese may believe that an adversary would not be willing to rely on a no-first-use pledge and would be deterred from a conventional attack on China.[4]

---

[2]With respect to the 1958 crisis, the Chinese complained that the Soviet nuclear "umbrella" had been extended over China only *after* the danger of nuclear confrontation with the United States had passed.

> The Soviet leaders expressed their support for China on September 7 and 19 [1958] respectively. Although at that time the situation in the Taiwan Strait was tense, there was no possibility that nuclear war would break out and no need for the Soviet Union to support China with its nuclear weapons. *It was only when they were clear that this was the situation that the Soviet leaders expressed their support for China* .... (Gittings, 1968, p. 92; emphasis added.)

Gittings cites "Statement by the Spokesman of the Chinese Government—a comment on the Soviet Government's statement of August 21" (1963) as his own source.

[3]Mao Zedong's outrageous statements to the effect that China could take in stride the loss of half its population to nuclear attack were part and parcel of the same strategy: China could absorb and eventually surmount even the worst that the United States (or the Soviet Union) could do to it.

[4]For example, if, as reported in Tyler (1996), a Chinese official delivered to a quasi-official U.S. visitor an implicit nuclear threat against Los Angeles (i.e., that China could use force against Taiwan with impunity because American leaders "care more about

If the Chinese now feel that their strategic nuclear force gives them a better deterrent than "people's war" provided, that would imply a greater willingness to be provocative and take risks. This is offset by China's greater interest in its relations with the United States and other Western countries, as a result of its current strategy for economic and technological development (as discussed above). In general, it does not appear possible to predict which effect will predominate with respect to Chinese behavior.[5]

USE OF FORCE AGAINST A WEAKER POWER

In looking at possible future Chinese uses of force, one can categorize them according to whether the target is, generally speaking, militarily stronger than China (or is the client of a stronger power) or not.

The Chinese uses of force in the South China Sea have fallen into the second category: The other claimants against whom China might use force (so far, Vietnam and the Philippines; potentially, Malaysia and Brunei)[6] are far weaker than China. While the Philippines has a defense treaty with the United States, that agreement does not cover the Philippines' claims in the South China Sea.[7] Thus, whether the United States would be seen as a patron of the Philippines in the case

---

Los Angeles than they do about Taiwan"), he would have been relying on U.S. skepticism concerning China's "no-first-use" pledge. Alastair Iain Johnston has argued that China is moving toward a "limited deterrence" nuclear strategy, in which nuclear weapons would be used to deter, *inter alia*, a conventional attack; this would appear to be incompatible with a no-first-use pledge. (Johnston, 1995/96, p. 19.)

[5]Indeed, it may not be entirely clear even after the fact: Was the 1996 military exercise, which included launching ballistic missiles on a trajectory that took them *over* Taipei to an impact point northeast of the Taiwanese port of Keelung, less provocative than the 1954 or 1958 shelling of Jinmen and Mazu?

[6]Taiwan occupies one island in the Spratlys, Itu Aba (Taiping Dao), but whether the PRC would have any interest in taking it remains unclear. Politically, removing this symbol of Taiwan's "Chineseness" would seem unwise. On the other hand, Itu Aba is the largest island in the Spratlys, and the Republic of China has soldiers stationed there; hence, the PRC might be motivated to take it because of its potential usefulness as a military base. (Valencia, 1995, pp. 6, 39.)

[7]The U.S. position is that the treaty does not cover the Philippines-claimed islands in the South China Sea, since the Philippines advanced its claims after the treaty was concluded in 1951. However, during a visit in August 1998, there were reports that U.S. Secretary of Defense William Cohen implied that the U.S. commitment might extend to Philippine forces located in the South China Sea. (Deocadiz, 1998.) It is not clear whether or not this represented a change of policy.

of a conflict with China would depend on the specific circumstances; as noted, at the time of the Chinese occupation of Mischief Reef in 1995, U.S.-Philippine relations were still frayed as a result of the non-renewal of the leases on U.S. military bases.

Future Chinese behavior toward the South China Sea can be expected, of course, to depend on a variety of factors. One can expect, however, that if China uses force in the future in this regard, it will pick an opportune moment when its intended target is relatively isolated from sources of outside support. This may be somewhat harder to do in the immediate future than in the past. Vietnam is no longer as isolated from its neighbors and the West as it was in 1988. And, in reaction to the Mischief Reef incident, the Philippines sought a greater measure of Association of Southeast Asian Nations (ASEAN) solidarity on the issue.[8] As for the longer term, China's ability to maneuver in this regard will presumably depend on relations among the ASEAN members and the strength of their ties to outside powers, primarily the United States.

China might also use outstanding border disputes to pressure neighbors, such as Vietnam and India, if larger issues made that course seem advantageous. Since China's concerns about Vietnam stemmed fundamentally from the latter's Soviet connection and its use of Soviet backing to seek a position of (as the Chinese put it) "regional hegemony," China will probably have less incentive to pressure Vietnam in the future. India, however, is another matter: India's bid for "regional hegemony" (the same charge that was levied against Vietnam in 1978–1979)—which is the interpretation the Chinese tended to put on its nuclear tests—and the possibility of perceived Indian involvement in any future unrest in Tibet could lead China to believe that it had to "teach India a lesson." In either case, one could envisage a limited Chinese border action, intended to produce a psychological shock and a change in policy.

Similarly, China might, hypothetically, attempt an "educational" campaign against a Central Asian state if it perceived that state as

---

[8]Of course, the prospects for this solidarity are presumably limited by the existence of overlapping claims by the ASEAN members themselves; the most significant deterrent to further Chinese activism in the region might be an agreement among the ASEAN claimants sorting out their conflicting claims.

supporting separatist groups in Xinjiang (or against Mongolia, if it were perceived as supporting separatism in Inner Mongolia).[9]

For the most part, the Chinese use of force against weaker states has not differed that much from the use of force against stronger states (or states with stronger patrons). The main exception has been the Chinese use of force in the South China Sea, where the goal has been to control islands and construct infrastructure (for the ultimate purpose of vindicating its territorial claims) rather than to achieve an objective by psychological or political means. This could presumably change as China grows stronger; however, with the possible exception of sparsely populated Mongolia, it seems difficult to envisage the Chinese embarking on a war of conquest against a weaker, isolated power in the next two decades.

USE OF FORCE AGAINST A STRONGER POWER OR THE CLIENT OF A STRONGER POWER

Obviously, if one is going to use force against a stronger power, or against the client of a stronger power, one must look for some particular advantage that can make up for one's overall relative weakness. Judging from the historical record, one can come up with a list of tactics, such as the following:

- use of surprise to create psychological shock

- inflicting casualties to create political pressure on the opponent

- creation of tension to divide the opposing alliance or to create political problems for an opponent

- creation of a *fait accompli*, presenting the opponent with a choice between acquiescence and escalation.

The PRC has used all of these methods in its conflicts with stronger powers. Chinese tactics in Korea used surprise to catch U.S.–United Nations troops in an unfavorable position and inflict a major defeat on them. The psychological shock thus created—which was so great that, at one point, the evacuation of the entire Korean peninsula was

---

[9]These possibilities are totally speculative at this point; no neighbor of China has shown an interest in challenging it in this way.

under consideration—affected U.S. behavior for the rest of the war. Once the battle lines stabilized, China conducted a war of attrition that attempted to place domestic political pressure on the U.S. administration to end the war. Alternatively, if the purpose of the military activity is to create tension without bringing matters to a head—as it appeared to be in the case of the Sino-Soviet border incident of March 2, 1969—China may attempt to keep the level of violence below the threshold that might prompt an escalation by the opposing side.

Chinese pressure against the offshore islands of Jinmen and Mazu was in part designed to divide Taiwan and the United States by playing on Taiwanese fears that the United States would be inclined to abandon the islands rather than face the prospect of a major war over such small stakes. At the same time, a steadfast U.S. administration could expect political trouble, both with respect to its own public and its major allies in Europe, who feared that U.S. involvement in Asia would weaken its ability to contribute to their defense against the Soviet Union.

Finally, China could hope to present a stronger opponent with a *fait accompli* that could only be overturned by escalating to a level of violence that might seem inappropriate to the stake involved in the conflict. For example, when, in early 1955, the PRC took the Republic of China (Taiwan)–held island of Yijiangshan, near the Dachen group (off the Chinese coast, several hundred miles north of Taiwan), the cost involved in recovering it would have been disproportionate to the value of the island itself. As a result, under U.S. pressure, the Taiwanese abandoned the entire Dachen group, which the PLA then occupied.

CREATING A CRISIS WITH RESPECT TO A STRONGER POWER

The historical record contains important instances in which China was willing to create a crisis even though it was militarily inferior to its potential opponent, i.e., the United States (in the Taiwan Strait crises of 1954–1955 and 1958) and the Soviet Union (with respect to the border incidents in 1969). Presumably, China could do this again (against the United States) if it felt that the United States was pursuing a policy that eventually would seriously threaten Chinese interests but that might still be alterable, if the United States could be convinced of its dangers.

For example, depending on the circumstances, China might see a future *rapprochement* between Vietnam and the United States as threatening, especially if it believed that a likely result would be regular U.S. military access to bases in Vietnam. To derail this development, China might seek to create a crisis between itself and Vietnam; it could, for example, exploit its territorial disagreements with Vietnam (both the land border and the maritime border in the Gulf of Tonkin, to say nothing of the overlapping claims in the South China Sea) toward this end. The purpose would be to cause the United States to believe that *rapprochement* with Vietnam would not be an easy, risk-free way of increasing its ability to bring military pressure to bear on China but instead could embroil the United States in a conflict in which it had little real interest and for which domestic political support might be very hard to obtain.

Chapter Five

LOCAL WAR UNDER HIGH-TECH CONDITIONS

The previous chapter dealt with the political-military context that both motivates the "local war under high-tech conditions" strategy and elucidates the political constraints under which it would be implemented. We turn now to the operational demands the strategy imposes on the PLA, which differ considerably from those the "people's war" strategy imposed. As this chapter shows, the PLA is far from being able to meet these demands in the fullest sense; in the following chapter, we discuss the consequence of this inability: the "asymmetric" tactics to which the PLA could resort in a conflict with an adversary force more fully able to meet the operational demands of this style of warfare.

The limited objectives and duration of the conflict greatly increase the importance of achieving a decisive result quickly, at the outset of the conflict. Thus, China's vast size and population, crucial assets in the traditional "people's war" concept, are, at best, irrelevant under local war conditions. The commander of one of China's seven military regions (Chengdu) highlights the contrast between the local war and "people's war" strategies:

> In the past, because we mainly based ourselves on total war in which the whole nation engaged the enemy forces, we employed a ... large-scale mobilization. In the future the main threat of war that we face will be local war and armed conflict under high-tech conditions. Because the war's objective will be limited, its element of surprise will be increased; the demand will be for mobilization to be partial and rapid. Whether one can react swiftly when war breaks out will directly affect the course and outcome of the war. (Liao, 1997.)

The necessity of striking quickly and the limited duration of local conflicts places effective limits on the number of troops that can be employed in a given local war and renders full-scale mobilization for war unnecessary if not obsolete.

The increasingly important role of high technology in modern warfare has been a prominent component of China's evolving military strategy since the late 1970s. Chinese strategists noted the influence that improvements in the range, precision, and destructiveness of weapon systems and in reconnaissance and communication capabilities were having on how warfare could be conducted. As early as 1982, Israel demonstrated the effectiveness and lethality of modern weapons in its rapid and decisive defeat of Syrian forces in the Bekka Valley. (Liu Fengcheng, 1995.) High-tech, modern weaponry enhances the speed and precision with which a country can apply military force, thereby increasing the ability of the possessor of such weapons to gain a decisive advantage over its opponents during the initial stages of a conflict. (Jiao and Xiao, 1987.) These qualities further intensified the need to be able to employ military force rapidly, which is already at a premium given the nature of local warfare.

Information, or information warfare, holds a central role in Chinese strategists' discussion of high-tech warfare.[1] China's conception of information warfare has two aspects. The first derives from the dramatic advances in intelligence and command and control systems. Improved sensor, communication, and guidance technologies allow an advanced military to gather and utilize information about the battlefield more rapidly and effectively than possible before. The result, according to an article in the *Liberation Army Daily*, is that:

> Today, the key to gaining the upper hand on the battlefield is no longer mainly dependent on who has the stronger firepower, but instead depends on which side discovers its enemy first, responds faster than the latter and strikes more precisely than the latter. (Wang and Lin, 1998.)

The challenge to any military engaged in a high-tech conflict, therefore, is to reduce one's adversary's ability to obtain or utilize infor-

---

[1]Interview with James Mulvenon, RAND.

mation and thereby degrade the adversary's overall effectiveness and, at the same time, to maintain the integrity of one's own command, control, and intelligence systems.

The second aspect of the Chinese view of information warfare is more exotic. It emphasizes the potential for staging "soft" attacks, against an adversary's information or computer system as the basis of an asymmetric strategy, particularly against an enemy with greater conventional military capabilities. The purpose of these attacks is to "infiltrate into and sabotage the other party's military information systems" to weaken its "command and control efficiency or paralyze the other party's command and control system." (Zhi and Liang, 1996.) There are a number of ways to attack an enemy's computer network, including the use of computer viruses or logic bombs. The Chinese may find conducting information warfare through cyberspace attractive because it grants them power-projection capabilities far beyond those of their conventional military forces and also affords China some degree of deniability. (Mulvenon, 1999, pp. 175–176.)

Chinese strategists had emphasized the growing importance of high technology in war since the late 1970s. They were, nonetheless, deeply impressed by the U.S. performance in the Gulf War. The 1995 Director of China's Commission on Science, Technology, and Industry for National Defense noted that:

> The Gulf War demonstrated that the application of high-tech in the military has given weapons an unprecedented degree of precision and power, heightening the suddenness, three-dimensionality, mobility, rapidity, and depth of modern warfare. The party that enjoys a high-tech edge clearly can exercise more initiative on the battlefield. (Ding, 1995.)

The devastation of Iraq's Soviet-design equipment highlighted the lethality of high-tech weapons and their ability to bring a conflict to a rapid conclusion. It also demonstrated the vulnerability of China's own weapon systems, most of which are also of Soviet design.

It is important to note, however, that the American performance against Iraq reinforced rather than redefined Chinese strategic thinking on the nature of war in the post–Cold War era. The U.S. war effort was surprising in the effectiveness of its operational execution, but the theoretical and strategic concepts it reflected did not come as a complete surprise to the Chinese.[2] One of the foremost academic observers of the Chinese military, Paul Godwin, contends that the PLA was at least "well prepared conceptually to respond to the technological challenge presented by the Gulf War." (Godwin, 1996, p. 473.) According to another scholar, the Gulf War "served to eliminate the lingering doubts among Chinese strategic planners on introducing the new local war doctrine, principles and tactics." (Li, Nan, 1996, p. 456.) He further notes that the Gulf War

> demonstrated that the resolution of high-tech local war tends to be quick and lethal, hence validating the principle of "fighting a quick battle to gain quick solution" and showing the undesirability, even impossibility, of fighting a protracted war of attrition under the conditions of modern, limited war. (Li, Nan, 1996, pp. 457–458.)

The challenge confronting Chinese military planners is to develop strategies and tactics that reflect the operational requirements of fighting a "local war under high-tech conditions" yet do not exceed the PLA's existing operational and technological capabilities.

REFORM AND MODERNIZATION

China has instituted a number of important military reforms since 1985 in response to the new operational demands of fighting a "local war under high-tech conditions." The fundamental aim of these reforms is to move the PLA away from its traditional large, unwieldy force structure to a smaller, more efficient one.[3] As described by one general, modernization is intended to change China's "armed forces construction from following a quantity and size model into following a quality and effectiveness model." (Zhao Nanqi, 1996.) The PLA made a significant move in this direction in 1985, when the CMC

---

[2]Varying alternative views are presented by David Shambaugh and Michael Pillsbury. Shambaugh describes the Chinese as "stunned" by the U.S. performance in the Gulf and contends that it caused a "thoroughgoing revision of doctrine and training in the PLA." (Shambaugh, 1996, pp. 25, 26.) Pillsbury, on the other hand, argues that many Chinese analysts saw the Gulf War as revealing primarily the weaknesses, rather than the strengths, of the U.S. military. (Pillsbury, 1998, p. 15.) The Chinese view of U.S. vulnerabilities is discussed in the next chapter.

[3]On this reform project, see the discussion of the "two transformations" in Finkelstein (1999), pp. 135–138.

announced troop reductions of 1 million personnel, a 25-percent reduction in overall manpower. Perhaps more significant politically, the number of PLA senior officers, a group generally less disposed to support the ongoing military reforms, was reduced the same year by a full 50 percent. (Baum, 1993, p. 379.) During the Ninth National People's Congress in March 1998, Li Peng announced a further reduction of 500,000 troops to be completed within three years.[4] These reductions, assuming they occur in full, will lower PLA troop levels to approximately 2.5 million early next decade, down from 4 million in early 1985.

The PLA has made significant changes to its overall regional structure and organization. In 1983, its 35 field armies were reorganized into 24 group armies, which integrated infantry and armor to facilitate combined arms operations. This change actually came two years before the move to the local war strategy. In 1985, China's 11 military regions were reorganized into seven. The individual military regions vary tremendously in terrain and climate and confront distinct security threats. As a result, each region develops and implements the training and field exercises that are most relevant to its particular environment, although cross-training among regions does occur where deemed appropriate. (Godwin, 1996, p. 467.) As one article explained, "to cope with modern limited war, it would suffice to use the forces of one military district, or at most the forces of two or three military districts." (Jiao and Xiao, 1987, p. 50.) PLA leaders viewed allowing each military region to develop capabilities to address the demands of its own, distinctive environment as more efficient than having a single, centrally directed force try to prepare for all possible contingencies that might erupt along China's long borders.

The development of elite units within each military region is an additional component of the PLA's program to enhance its ability to respond quickly to military crises on its periphery. These units, called *fist troops*, are rapid-reaction forces organized in battalion- or brigade-sized units.[5] Modeled on rapid mobility units in the West, such as the U.S. Army's 82nd Airborne and 101st Air-Assault divi-

---

[4]"Li Peng's NPC Report: Military Cuts on Schedule" (1998).

[5]Concerns about internal stability imply a requirement to be able also to transport troops quickly within the country.

sions, the fist troops are intended to be able to deploy anywhere in China within 12 hours. (Godwin and Schultz, 1993, p. 4.) The use of such forces is new to the PLA but features prominently in the writings on local war appearing in Chinese military periodicals. These troops are viewed as vital to China's ability to meet security challenges in the near term on China's borders, either through actual military conflict or through deterrence. (Jiao and Xiao, 1987, p. 50.)[6]

The role of elite units in the PLA highlights the dramatic contrast between the requirements of the newer military strategy and traditional "people's war." Under "people's war," victory was achieved through the efforts of the entire nation, whose political commitment to the struggle and ideological fervor were expected to make up for whatever the nation lacked in military materiel and expertise. Successfully fighting a "local war under high-tech conditions" calls for the opposite: small numbers of highly trained, professional soldiers who are able to respond swiftly to any potential crisis.

The move away from Mao's "people's war" concept also instigated significant changes in the way the PLA conducts its training. The broad goals of these changes were to improve the professional and technical quality of PLA personnel, improve the effectiveness of PLA training methods, and improve the PLA's capacity to conduct joint and combined-arms operations. While far from discounting the need to maintain the PLA's strict allegiance to the CCP's authority and ideology, military reforms placed much greater emphasis on ensuring PLA cadres had sufficient military and technological expertise. In an article appearing in a 1979 issue in the leading CCP journal *Hongqi*,[7] PLA Marshall Xu Xiangqian stated that the PLA "must train large groups of red and expert cadres who have a good knowledge of modern science and technology and are capable of commanding joint operations in a modern war."[8] Six years later, in an

---

[6]A report in the summer of 1999 noted that a "special forces combat unit" of the Guangzhou Military Region was expert in "seizing beaches by stealth from the sea." (Huang Rifei, 1999.) The emphasis on amphibious assault was presumably connected to Chinese displeasure with Taiwanese President Li Teng-hui's call for "special state-to-state" relations between China and Taiwan.

[7]Now known as *Qiushi* [Seeking Truth]. *Hongqi* means *red flag*.

[8]It is important to note that even stating that it is *as* important to be "expert" as "red" is a break from the Mao era, when political factors (i.e., being "red") were considered far more important than technical competence. (Xu, 1979, p. L17.)

article in the same publication, PLA Chief of Staff Yang Dezhi declared the development of capable military personnel as the "key" to army modernization. He argued,

> If we [the PLA] do not have people who have a good grasp of modern science, culture, and technical skills, it will be impossible to bring into full play the power of modern weapons. (Yang, 1985.)

Chinese military training has grown increasingly sophisticated since the end of the Gulf War (although it is difficult to determine to what extent training objectives have in fact been achieved). PLA training exercises focus on developing and improving its ability to conduct joint operations under local war conditions. They are intended to produce realistic, competitive scenarios in which the units involved can enhance their rapid reaction capability, nighttime operation capability, and proficiency with new, high-tech weaponry and equipment.[9]

In light of its resource constraints, PLA modernization is proceeding in a gradual, but focused, way. As Liu Huaqing explained, "We must proceed from our country's conditions and cannot compare everything with advanced international standards, nor pursue unrealistically high indexes and high speed." (Liu Huaqing, 1993, pp. 19–20.) Through domestic development and foreign acquisition, the PLA is selectively upgrading its weaponry and equipment in a pattern consistent with the operational demands of "local war under high-tech conditions." The PLA Air Force (PLAAF) and PLA Navy (PLAN) have, thus far, been granted priority. While infantry and armor units have received less attention, significant resources have been directed toward ballistic and cruise missile programs. (Liu Huaqing, 1993, p. 21; see also Khalilzad et al., 1999, Ch. 3.)

Some progress has been made in improving the PLA's arsenal in limited areas. The PLAN has commissioned two new, indigenously produced, classes of surface combatants in the 1990s: the 4,500-ton *Luhu*-class guided missile destroyer and the 2,750-ton *Jiangwei*-class guided missile frigate. In addition, China is building the larger (over 6,000 tons) *Luhai*-class destroyer, to be equipped with C802 ship-to-

---

[9]"Media Report: Cross-Regional Training" (1995). See also Blasko, Klapakis, and Corbett (1996).

ship missiles with a range of 120 km.[10] China's navy has further added to its forces through the purchase of two Russian-built *Sovremenny*-class destroyers and at least four *Kilo*-class diesel submarines. The PLAAF held the first test flight for its much-anticipated F-10 multirole fighter in March 1998. The F-10, China's first indigenously designed "fourth-generation" fighter, has performance capabilities roughly comparable to the Lockheed Martin F-16A and is a dramatic improvement over China's existing inventories.[11] Like the PLAN, the PLAAF has also improved its equipment through the acquisition of higher-performance Russian equipment, purchasing Su-27 air-superiority fighters and securing a production license to construct over 200 more in China. China also enhanced its air-defense capabilities through the acquisition from Russia of eight batteries of the SA-10 anti-air missile system. Finally, the PLA demonstrated the accuracy of its indigenously produced Dongfeng-15 short-range ballistic missile during its missile tests off Taiwan in 1995 and 1996.

China has actively sought to acquire foreign weapon systems and technologies, and the PLA's shopping list has been varied.[12] Nonetheless, it is worth noting that, thus far, Beijing does not appear to be in the process of comprehensively upgrading the PLA's equipment by purchasing foreign systems and technologies. As Liu Huaqing notes, "A big developing socialist country like ours cannot buy modernization of the whole army." (Liu Huaqing, 1993, p. 19.) China has twice suffered as a result of seeking outside assistance in developing military technology, at the hands of the Soviets in the 1950s and of Western suppliers in the 1980s, and therefore does not want to become dependent on foreign suppliers again. Its strategy is to acquire advanced foreign systems and technologies for a narrow range of platforms, while relying on upgrading indigenous production capabilities for longer-term development. Again, as Liu explains,

> One of the basic principles of modernization of weapons and equipment in our army is to rely mainly on our own strength for

---

[10]Periscope, *Daily Defense News* (Internet edition), January 20, 1999.

[11]"First flight for F-10 paves way for production" (1998).

[12]See Khalilzad et al. (1999), pp. 50–53, for a brief discussion of Chinese foreign arms purchases.

regeneration, while selectively importing advanced technology from abroad, centering on some areas. (Liu Huaqing, 1993, p. 19.)

CHALLENGES THE PLA CONFRONTS

China's military leaders are well aware that the PLA remains years, if not decades, away from being able to fight a "local war under high-tech conditions" as the Israeli, British, and American militaries fought them in the 1980s and 1990s. As Godwin puts it, the Gulf War

> served primarily to underscore what they [the PLA leadership] had known for many years: that the absence of modern armaments severely restricted the PLA's effectiveness, and could even endanger military success under the restraints provided by limited war's requirements for speed and lethality in combined arms warfare. (Godwin, 1996, p. 473.)

In virtually every area, the PLA's technological capabilities lag far behind those of not only the United States but many of China's regional neighbors, such as Japan and Russia.

Many articles appearing in Chinese publications are very frank regarding the challenges the PLA faces as it attempts to develop the operational capabilities required to fight a "local war under high-tech conditions." One of the most serious of these challenges results from the multitude of problems that afflict the PLA as an institution. General Zhao Nanqi, director of the China Military Science Association (and former head of the PLA General Logistics Department), describes the PLA as "characterized by a large size, backward weapons and equipment, swollen bureaucracies, irrational regulations and personnel schemes, and a low quality of personnel." As a result, China's armed forces find themselves in a "hard-to-rectify situation of accumulated difficulties." (Zhao Nanqi, 1996.) Zhao's criticism highlights three primary areas of difficulties for the PLA: (1) poor hardware, (2) insufficiently skilled personnel, and (3) institutional inertia or outright resistance to change.

The PLA's most obvious flaw is that much of its equipment is obsolete. The bulk of China's land, air, and sea forces are made up of antiquated platforms in 1950s- or 1960s-era Soviet technology. In light of this fact, many Chinese analysts worry that the growing importance of high-tech systems in modern warfare will seriously challenge China's ability to defend its national interests in the future. (See Zhao Nanqi, 1996; Shi Zhongcai, 1996; and Chen, 1997.) A 1987 article in *Jiefangjun Bao* noted that the PLA's technological level lagged far behind "world advanced levels" and warned that the failure to address this gap could result in "great losses in future wars." (Pan, 1987.)

To many Chinese military thinkers, the low quality of PLA personnel is of as great, if not greater, concern than its outdated weapon systems and equipment. Warfare in the modern era is becoming increasingly complex in terms of the kinds of equipment and technologies individual soldiers must operate and the nature of the operations and tactics the soldiers must use to be successful. However, serious problems continue to exist in the technical and professional capabilities of PLA personnel. In a 1996 article, a Chinese military journal lamented, "Compared to other military forces, our military currently has the problem of 'technological inferiority' in equipment and the problem of 'knowledge inferiority' in the quality of manpower." The article warned that China's inability to improve the technical and professional caliber of its personnel might call its ability to utilize military force as an effective tool of national policy into serious question. (Hong and Tian, 1996.)

The PLA has had some institutional difficulties adjusting to its newer military strategy. The PLA has traditionally played an important political and civic, as well as security, role within the Chinese state. However, China's military reforms since 1978 have placed unprecedented emphasis on developing the PLA's professional military qualities. Given the PLA's sheer size and past emphasis on political and other nonmilitary criteria in managing personnel and other organizational matters, it should be no surprise that institutional reforms generated by changes in doctrine have encountered difficulties. An article in *Chinese Military Science* noted that the PLA has "yet to solve" the problem of "freeing itself from the influence of old conventions, familiar rules, and outdated concepts." (Huang and Zuo, 1996.) A PLA political officer made a more focused criticism of PLA personnel management in a 1996 article appearing in the journal *Guofang* [*National Defense*] in which he noted that "there are still instances (within the PLA) in which competent personnel are suppressed and their services are not put to good use." (Shi

Zhongcai, 1996.) The article claims that factionalism and valuing seniority over ability remain serious problems within the PLA.

There is evidence in the Chinese media of continued resistance within the PLA to the implementation of the military reform and modernization programs. This evidence rarely takes the form of explicit criticism of the "local war under high-tech conditions" strategy. Rather, it is most frequently found in the writings of supporters of the local war model, who complain that the modernization process is being impeded by those who do not share their beliefs. One such article from *Chinese Military Science* noted that armies frequently retain operational doctrines that had proved effective in past conflicts and use them in their preparation for future conflicts. Such an army will not realize the error of its ways until it finds itself at a disadvantage in the next war. "So, in this sense, the victory of a previous war has sown the seed of failure of the next war." The authors maintain that this demonstrates the need to "build an operational doctrine" to enable the PLA to compete in a future, high-tech war. (Huang and Zuo, 1996.) The sentiments expressed in the above quote are a warning that the PLA will face certain failure in its next conflict if it cannot move beyond the "people's war" tactics that it believes were so successful in past wars.

Occasionally, explicit criticism of military reform appears in print. One such article, in an April 1998 issue of *Liberation Army Daily*, argued that "people's war" theory "should remain as a basic requirement for the PLA combat theory research." The article notes that the Gulf War created great concern among Chinese strategists over the dangers of high-tech war and the relevance of "people's war" to China's modern security needs. However, according to the article, most of these concerns were "dispelled" following the intense study of high-tech weaponry and high-tech local war and ensuing "theoretical and ideological debates." The article asserts that the few people who continue to turn "a blind eye" to the "people's war" theory are guilty of blindly worshipping western combat theory and belittling the PLA's "people's war" tradition. (Huang Jialun, 1998, p. 6.) This article demonstrates that individuals or groups remain within the Chinese military establishment that oppose the PLA's adoption of the "local war under high-tech conditions" strategic concept.

In addition to those described above, the PLA is confronting at least two other significant challenges that do not stem directly from its own institutional weaknesses. The first is simply that the PLA lacks experience conducting the kinds of complex operations they believe are necessary to win a "local war under high-tech conditions." The kinds of combined and joint operations that their new military strategy calls for are dramatic departures from the PLA's traditional warfighting methods and strategies. Training exercises are being developed to improve the PLA's capabilities in such areas as its ability to mobilize rapidly, conduct joint or combined operations, conduct night combat and mobile operations, and conduct electronic warfare.[13] As the PLA establishes and expands these exercises, its operational capabilities will no doubt improve. Nonetheless, it remains to be seen how rapidly these capabilities will improve and whether they will be sufficient for the PLA to be successful in its next armed conflict.

The final challenge PLA's modernization confronts is financial. While the PLA's budget has risen significantly during the 1990s, Beijing does not appear to be undertaking a "crash" program to complete military modernization in the short term.[14] This reflects the Chinese leadership's determination that national defense modernization should be subordinate to the economic and technological modernization of the country. As an article in the *People's Daily* [*Renmin Ribao*] explained, "If divorced from the center of economic construction, the modernization of national defense may lose its material basis. Therefore, the army should exercise restraint." (Li Weixing, 1996.) The PRC's unwillingness to devote more resources toward military modernization also implies that its leaders believe the prospects for significant military conflict in the immediate future are relatively low.[15]

---

[13]For details, see Zhang Guoyu (1995), Wang Mingjin (1996), and Liao Xilong (1997).

[14]For more detail on China's defense budget, see Khalilzad et al. (1999), pp. 37–39.

[15]As of this writing, it is too early to tell if the NATO intervention in Kosovo has significantly influenced Chinese thinking in this regard.

THREAT PERCEPTIONS

The "local war" concept does not provide any formulaic criteria that deterministically identify China's most likely adversaries in future wars. While certain kinds of behavior on the part of other countries can raise or lower the likelihood of military conflict, China no longer believes, on the basis of ideology, that any country *intrinsically*, i.e., due to its imperialist (late capitalist) or social-imperialist nature, poses a threat to Chinese security.[16] Military conflict can occur suddenly for a myriad of reasons ranging from territorial disputes to ethnic or religious strife. (Shi Yukun, 1995.) Its occurrence is neither inevitable nor predictable.

It is rare that Chinese media explicitly name individual countries as threats to China's security. Nevertheless, the descriptions of countries that are said to pose a potential threat to Chinese interests, although vague, are generally sufficient to provide the reader with a fairly good idea of the country or countries to which the article is referring.

The country that is most frequently identified, either explicitly or implicitly, as a possible military adversary of China is the United States. The United States is the strongest military power in East Asia. While the two countries have no specific territorial disputes, Chinese and U.S. interests have the potential to conflict in areas ranging from the Korean Peninsula to the South China Sea. As the heightened tensions in March 1996 demonstrated, nowhere is the potential for military conflict between China and the United States greater than in Taiwan. Moreover, the Clinton administration's dispatching of two aircraft carrier battle groups to the vicinity of Taiwan dramatically reminded Chinese leaders of the decisive role the U.S. military can play in a conflict there. Similarly, the U.S.-led NATO intervention in Kosovo, on human rights grounds, set an unwelcome precedent.

In addition to the potential military threat, the Chinese have viewed a number of U.S. actions during the 1990s—such as threatening sanctions over human rights violations, opposing China's achieve-

---

[16]But see Christensen (1996a) for the argument that Chinese distrust of Japan is "historically rooted and visceral" (p. 41) and that Chinese analysts believe Japanese military renaissance would be inherently threatening to China, almost regardless of the overall international situation (pp. 40–45).

ment of "developing nation" status with respect to the World Trade Organization, and allowing Taiwanese president Lee Teng-hui to visit the United States—as evidence of Washington's desire to undermine China's domestic stability, economic development, international standing, and goal of national unification.[17] As one Beijing journal asserted, these kinds of actions "are essentially all aimed at opposing and eventually causing the collapse of socialist China through the peaceful evolution of China." (Hong Ye, 1995.)[18]

With respect to non-U.S. adversaries, Chinese military journals portray the PRC's most plausible security problems as originating from beyond China's maritime, rather than continental, border. Obviously, Taiwan looms large in Beijing's military planning. The PRC has consistently ruled out any commitment not to use force against Taiwan. Until the situation with Taiwan is resolved, China will have to devote significant resources toward planning and preparing for a military contingency, perhaps involving the United States, in the Taiwan Strait.

More generally, many Chinese analysts believe that the countries of maritime Asia are placing increased importance on securing or even expanding their "maritime rights and interests," often at China's expense. These analysts view the oceans as "the new high ground of international strategic competition" in the post–Cold War security environment. (Yan and Chen, 1997.) Not surprisingly, many of the articles that argue most forcefully that China should improve its ability to defend its maritime interests were written by PLAN officers or appear in the naval publication *Jianchuan Zhishi* [*Naval and Merchant Ships*]. (Yan and Chen, 1997; Li Yaqiang, 1995, and Shen, Zhou, and Zhan, 1995.) Some of these arguments, therefore, likely represent special pleading by members of the PLAN who would benefit if the navy received more resources and greater prominence within China's overall security strategy. However, it is important to note that the PLAN's arguments do resonate, to some degree, within China's higher-level leadership. In his 1992 work on China's strategy in the South China Sea, John Garver notes that China's "paramount

---

[17]Beijing has also been concerned that the United States might provide Taiwan with new theater missile defense capabilities.

[18]The U.S.-led NATO intervention in Kosovo, in the name of human rights, played into this perception.

leaders" believe control of the South China Sea, and the resources that lie within it, is important to China's long-term economic security, as well as its ability to achieve global power status. (Garver, 1992.)

In terms of specific countries, these threat perceptions are directed toward Southeast Asian states that have disputes with China in the South China Sea. Chinese publications discuss "certain medium and small coastal countries" that may menace Chinese maritime sovereignty in the future. Such statements almost certainly refer to Vietnam and perhaps the Philippines. China clashed with Vietnam over control of various reefs and shoals in the Spratly Island group during the 1980s (and with the Philippines over China's occupation of Mischief Reef since 1995), and tensions over the countries' competing claims in the region remain high to this day.

Chinese leaders are deeply suspicious of Japan's long-term intentions. China and Japan hold conflicting claims over the Diaoyutai, or Senkaku, Islands in the East China Sea. China views Japan's recent expansion of its defense perimeter as potentially threatening to its interests. (Zheng and Zhang, 1996.) The 1997 revised defense guidelines of the U.S.-Japan alliance, which state that the alliance covers "situations in the areas surrounding Japan," also raised Chinese fears that Japan might become more involved militarily in a future crisis or conflict involving Taiwan. Indeed, Chinese analysts may see Japan as ultimately more interested than the United States in hindering Chinese attempts to incorporate Taiwan into the mainland.[19] Japanese participation with the United States in theater missile defense development has also ignited Chinese fears, either because it might appear to be the construction of a "shield" behind which Japan might develop offensive military capabilities or because a mobile (e.g., ship-based) or wide-area system could contribute to the defense of Taiwan. (Christensen, 1999, pp. 64, 66, 75–76.)

It is interesting to note that, if Chinese strategists do have security concerns with regard to China's land borders with Russia, the Central Asian Republics, or India, they do not write about them. China and

---

[19]This conclusion was drawn by Christensen (1999, p. 63) on the basis of interviews conducted from 1993 to 1998 with a number of Chinese academics and military and civilian analysts in government think tanks.

the states of the former Soviet Union along its border appear to have largely resolved the majority of their historically contentious boundary disputes. China and India have also made substantial progress over the past ten years in lowering tensions around their sizable territorial disputes. Even following India's recent nuclear tests, Chinese regional specialists appear to discount the threat India poses to Chinese security. Russia appears to be a slightly different case. Chinese writings do not discuss any future potential threat Russia may pose to the PRC. In discussions about this issue, however, Chinese security analysts indicated that there is some concern in China that, when Russia does emerge from its current economic and political problems, which the analysts expect in about 15 years, it may once again pose a serious security threat along China's northern border.[20]

---

[20]Discussions with Chinese regional analysts in 1998.

Chapter Six

APPLICATION OF THE STRATEGY: DEALING WITH THE UNITED STATES

The specter of military conflict with the United States raises very difficult problems for the PLA. The U.S. military excels in virtually every aspect of the operational capabilities Chinese strategists identify as being important to winning a "local war under high-tech conditions." The United States is described as "being at the vanguard" of the revolution in military affairs. (Wang, Su, and Zhang, 1997.) One Chinese general stated that the United States waged the Gulf War "with great strategy, great command, great logistics, and a great alliance forming a great system." He concluded that the U.S. performance represented a "big step forward in both military theory and practice." (Shi Yukun, 1995.)

There is little doubt that at least some leaders in the PLA are aware of the dangers they would face in the event of a military conflict with the United States. One article in the *Liberation Army Daily* notes that the United States possesses a wide range of weaponry—such as reconnaissance satellites, airborne early warning systems, and precision guided weapons—that China does not. Even in terms of systems or platforms that both China and the United States possess—such as tactical guided missiles; command, control, and communication (C^3) systems; fighter aircraft; and submarines—those belonging to the United States are far superior to China's. In the face of such a drastic technological disparity, some Chinese military thinkers readily admit that "the obstacles facing China on the battlefield will multiply dramatically." (Lu, 1996.)

Unfortunately, an awareness of the areas in which U.S. military capabilities surpass those of the PLA is not sufficient to compel China to renounce its current interests that have the potential to bring it into military conflict with the United States. An understanding of China's own vulnerability to U.S. military power has probably made Chinese leaders more cautious in how they consider using military force in pursuit of its interests, but it will not convince the leaders to relinquish them outright. Instead, the task falls to the PLA to develop a strategy for dealing with U.S. military power in the event the use of military force becomes necessary to pursue or protect important Chinese interests.

No single open-source document provides an authoritative description of the strategies and tactics the PLA would use in a conflict with the United States. There are, however, articles that discuss U.S. military vulnerabilities, its strategic or tactical objectives in limited conflicts, and errors that past opponents of the United States or of other nations with high-tech military capabilities have made. Chinese military journals also engage in a modest discussion of the kinds of measures that China will need to take to defeat an enemy of superior military-technological capabilities. From these sources, it is possible to construct a general outline of a potential Chinese strategy for successfully engaging U.S. armed forces in a limited military conflict and achieving Chinese operational and political objectives.

U.S. VULNERABILITIES

Despite the current military preeminence of the United States, Chinese analysts identify a number of military and political vulnerabilities that China might be able to exploit in the event of a Sino-U.S military conflict. These perceived American military vulnerabilities fall into two categories: those arising from the environment in which U.S. forces will have to engage Chinese forces and those arising from U.S. reliance on highly complex and sophisticated technologies and systems to wage war.

Although China might ideally wish to bog down the United States in a land war, the most plausible sources of military conflict between the United States and China fall along China's maritime periphery, and the chances of engagement between U.S. and Chinese land forces are relatively low. Nevertheless, a military conflict with China could require the United States to employ significant air and naval resources. Even with existing U.S. bases in Asia, Chinese analysts argue that concentrating sufficient firepower and logistics in such a distant area of the world will be a serious challenge to the U.S. military. The extremely high consumption rate of material in high-tech conflicts will only compound the challenges confronting American forces in this regard. The conclusion, one Chinese analyst has implied, is that U.S. logistics lines will represent relatively vulnerable and extremely lucrative targets. (Yu, 1996.)

The maritime environment along China's coast also presents at least a modest challenge to U.S. naval capabilities. The waters around Taiwan and the Spratly Islands, another area of potential Sino-U.S. conflict, are relatively shallow. An article in *Liberation Army Daily* contends that "most of the U.S. Navy's equipment cannot meet the needs of a coastal battle because they were originally designed for ocean-going battles." The article points to U.S. weaknesses in detecting and attacking conventional submarines in shallow waters and deficiencies in anti-mine capabilities as evidence supporting this claim. (Wang Jianhua, 1997.)

Chinese military analysts claim that U.S. high-tech information systems, on which they believe its forces are increasingly dependent, are relatively fragile and unreliable and therefore represent a weakness that can be exploited. (Yu, 1996.) The United States is more effective than any other military in the world in using highly sophisticated systems to gather, analyze, disseminate, and use information about an operational setting to maximize the overall effectiveness of its military power. A force whose units have available a common situational awareness (a "digitized" force) would be able to maneuver and fire with much greater efficiency. However, according to a researcher at China's Academy of Military Science,

> if digital [i.e., digitized] forces lose the power to control information, they will not be able to stand up to mechanized forces [such as the PLA's]. If they fail to acquire or transmit information, digital forces will be paralyzed, their combat capability would shrink rapidly, and they will lose the initiative on the battlefield. (Lin, 1995.)

Chinese strategists may view information warfare as a particularly attractive way to exploit the U.S. overreliance on high-tech command, control, communication, and intelligence (C^3I) systems.

Aside from these technical factors, many Chinese analysts believe that the United States has developed a strong cultural aversion to incurring casualties in military operations. As a result, the United States would seek to avoid committing military force in a situation in which heavy casualties are likely to occur. Furthermore, the United States would be reluctant to enter into military conflict against a country that has a "strong national will and whose armed forces and people have the tradition of being willing to sacrifice their lives." (Fu, 1997.)

The articles discussing the U.S. aversion to casualties do not describe exactly how China might inflict heavy losses on U.S. forces. Given the most likely scenarios for Sino-U.S. conflict, which would feature potentially intense air and naval, but not land, campaigns by U.S. forces, the possibilities for doing so appear somewhat limited. So long as there are not significant numbers of U.S. troops on the ground in the PRC, the PLA's only real targets would be U.S. naval vessels and bases in Asia. While the PLA's current ability to attack these targets successfully is limited, it will improve significantly over the next five to ten years as the PLA incorporates more sophisticated systems, such as the recently purchased Su-27s and *Sovremenny* platforms or its increasingly capable ballistic missile forces, into its force structure. Of course, China could also resort to nuclear, chemical, or biological weapons. Attacking the U.S. Navy or U.S. bases in Asia, to say nothing of the use of weapons of mass destruction, would bring great risks to Chinese security. It could provoke an even stronger U.S. commitment of forces to the theater or even bring other nations into the conflict against China. However, such an attack might make sense if Beijing believed it would demonstrate to Washington, and perhaps Tokyo, that the costs they would incur in a fight with China were significantly higher than what their populations would accept.

In the face of the U.S. vulnerabilities listed above, China believes it would enjoy some important advantages. Since Chinese forces would be able to engage U.S. forces operating near its own territory, the PLA would not face the problem of maintaining long logistics lines, as U.S. forces operating in Asia would. To the extent that naval confrontations occurred in shallow and "brown" (i.e., coastal) waters, the PLAN could attempt to exploit perceived U.S. weaknesses in mine sweeping, and poor acoustic conditions would provide its submarines some protection from superior U.S. antisubmarine warfare capabilities.

China would be able to rely on a "vast, stable rear area" from which operations can be launched or supplied. (Yu, 1996.) The U.S. military is certainly able to strike targets on the Chinese mainland, and many of the important targets would be relatively close to the coast. Nevertheless, China is a nation of great geographic size, and U.S. air power simply could not conduct "parallel warfare" (i.e., simultaneously attack a range of targets throughout the country, with the aim of achieving a paralyzing shock effect) against China as it did against Iraq during the Gulf War. Moreover, China's strategic nuclear forces would constrain the extent and nature of the air campaign against the Chinese mainland, both through their deterrent effect and because the United States might not wish to degrade China's strategic connectivity (i.e., the ability of the PRC leadership to communicate with its military forces, especially its nuclear forces, throughout China).

CHINESE MILITARY OPTIONS

The discussion in Chinese military journals of the strategies and tactics needed to defeat the United States in a military conflict fall into two general categories. The first involves ways to prevent the United States from accomplishing its objectives vis-à-vis the PLA. A senior colonel in the Academy of Military Science describes the generic objectives of the U.S. military in a conflict to be to "disrupt and upset the enemy's operational rhythm, undermine the enemy's operational systems at all levels, and deprive the enemy of its overall offensive and defensive capability." (Huang and Zuo, 1996.) To a large extent, the United States has succeeded in attaining these aims in each of the conflicts it has been involved in since the Vietnam War.

The U.S. ability to accomplish its goals of "disrupting" its enemy's operations is based on its ability to locate and accurately strike not only high-value weapon platforms, such as tanks or anti-air batteries, but also systems that facilitate the more effective use of such assets, i.e., C^3I systems. To avoid a fate similar to what befell Iraq in the Gulf War, Chinese strategists hope to increase significantly the difficulties U.S. military forces would encounter pursuing this goal against China.

The most obvious way to make it more difficult for the United States to destroy Chinese military assets is to make it more difficult for U.S.

forces to find them. An article in *Liberation Army Daily* states that the purpose of U.S. efforts to "digitize" the battlefield is to remove the "fog" of war. The article argues, however, that these efforts themselves serve as powerful incentives to other countries to develop ways to "create more fog." (Li Bingyuan, 1996.) Decoys and electronic countermeasures can be used to deceive or defeat U.S. attempts to locate and target Chinese assets.[1] The PLA's capabilities in this area are very real. China is one of the world's leaders in "obscurant" technology—technology designed to degrade an adversary's reconnaissance, surveillance, target acquisition, and weapon guidance capabilities.[2] To put it another way, the PLA excels in being able to create, quite literally, fog on the battlefield.

Should the PLA fail in its attempts to deny the United States knowledge of the location of certain platforms or systems, the assets must be able to withstand the inevitable U.S. assault. Chinese strategists observed during the Gulf War that deeply buried structures can be very difficult to destroy with conventional weapons. These kinds of defensive works can provide an important means of protecting Chinese assets from powerful and accurate U.S. missile and aircraft strikes.

Despite being relatively backward technologically, China's military relies on modern C^3 systems to process and disseminate information. To ensure survivability in the event of a Sino-U.S. military conflict, this system must be "extremely flexible, well-concealed, and able to operate under a wide variety of conditions." (Wang, Su, and Zhang, 1997.) That is, the PLA hopes to reduce the potential vulnerability of its C^3 systems by making them as robust as possible. The articles covering this topic are rather vague about how this is to be accomplished. One article called for the construction of multiple C^3I networks that use a number of frequency bands and power sources. (Wang, Su, and Zhang, 1997.) Another article, appearing in *National Defense*, admits that China's technological limitations will make communication and information security more difficult. It calls for

---

[1]Options like these can be found in the following articles: Lin (1995), Zhang Dejiu (1996), Jiang and Chen (1993), and Sun Zi'an (1995).

[2]Office of the Under Secretary of Defense for Acquisition & Technology (1996), pp. 15-2 and 15-21, discusses the effects of "obscurants" on the electromagnetic spectrum, e.g., ultraviolet, visible, infrared, millimeter wave, and microwave.

the development of a comprehensive system that incorporates military and civilian equipment with the goal, again, of creating as flexible and robust a system as possible. (Zhang Xiaojun, 1995.)

China can use a number of nontraditional methods to preserve the integrity of its communication systems. For example, an article published in 1995 suggests that China use the communication systems of nonbelligerent nations, arguing that an adversary would be unlikely to attack a third party's communication facilities. "Therefore," it concludes, "consideration should be given to using the circuits of these countries' satellites to send telegrams or commands to troops engaged in field operations in the war zone." (Sun Zi'an, 1995.)

The article also suggests that the PLA utilize unused civilian television or radio frequencies, making the somewhat less persuasive claim that "the enemy would risk provoking strong opposition in the world media if it interfered with or damaged" civilian systems. The PLA should also use "mobile or handheld telephone exchanges." According to the article, the bulk of the equipment involved can be deployed underground (obviously, there would have to be antennas as well) or moved around easily. It contends that they are easy to hide and highly survivable and, therefore, are a "fairly ideal means of backup communications." (Sun Zi'an, 1995.) As is typically the case in articles of this sort, the discussion does not provide enough detail to assess how carefully these options may have been investigated: For example, would these alternative systems provide sufficient bandwidth for military communication?

The second type of strategies and tactics that Chinese military journals discuss focuses on disrupting U.S. operations to prevent the timely deployment of significant U.S. forces into a given war zone. Indeed, this would be the central task confronting the PLA in a limited conflict with the United States. Chinese analysts recognize that the kinds of high-tech weaponry the United States possesses allow it to inflict heavy losses on an adversary in a very short time. (Wang Chunyin, 1997.) Colonel Yu Guohua, a lecturer at China's National Defense University, writes that the speed, accuracy, and destructiveness of high-tech weapons

> greatly increases war's abruptness and quick-decisions and often gives the disadvantaged side no breathing room. Sometimes even before the weaker side can respond or has a chance to fight back, the war is already over and victory is already decided. (Yu, 1996.)

Thus, allowing the United States time to build up its forces in the theater to launch an attack, as Iraq did, invites disaster.

As already noted, the Chinese believe that there will be strong international and domestic pressures to bring conflicts to a rapid conclusion and that, therefore, delaying the entry of U.S. forces into combat may be sufficient for China to achieve its political goals. Chinese strategists identify U.S. C^3I and logistics systems as key military assets to target. An article in *Chinese Military Science* [*Zhongguo Junshi Kexue*] describes modern warfare as "a rivalry between systems." By identifying and striking the most vulnerable "nodes" in the U.S. operational system, the article argues, the PLA could "destroy and paralyze" the U.S. operational structure. (Huang Xing and Zuo Quandian, 1996.) The U.S. dependence on highly sophisticated systems to conduct military operations increases the number of targets available. (Huang, Zhang, and Zhang, 1997.) As discussed earlier, Chinese strategists also view these systems as inherently fragile and susceptible to attacks at vulnerable points that can put the entire system at risk. (Yu, 1996.)

As noted in the Introduction, these strategies are discussed only in general terms in the open-source literature. Thus, it is difficult to tell whether these articles reflect a Chinese belief that they have solved these problems in operational terms or whether they are meant to set out a task for planners to address. In any case, the articles represent evidence that at least some Chinese strategists are thinking along these lines, which appear to be logically derived from the overall understanding of what "local war under high-tech conditions" involves.

The U.S. style of warfare consumes extremely large amounts of material over relatively short periods. The U.S. military, therefore, depends on stable and secure logistics systems to maintain a sufficient supply of material to continue its operations. As noted earlier, a number of Chinese analysts view these long and relatively vulnerable logistics lines as valuable targets. As one article put it,

> We [the PLA] have to take drastic measures—take away the firewood from under the caldron—and use every means to ruin the enemy's logistic protection system, so that the entire combat system loses its dynamics and vitality, miring the enemy in passivity where it must take a beating. (Yu, 1996.)

An article in the journal *Modern Weaponry* [*Xiandai Bingqi*] argues that cruise and ballistic missiles can provide China with the means to effectively interrupt U.S. logistics flows into East Asia and strike "strategic points or fleet [sic] of aircraft carriers."[3] The article contends that these missile systems are "easy to develop and inexpensive to make" and are able to "attack the enemy on the move within a certain range, thereby achieving the goal of blocking or delaying enemy operations or inflicting on it direct losses."[4] "Saturation strikes" of ballistic missiles are very difficult to defend against. Such attacks could be effective against "high priced air defense missile systems and other large-scale equipment." (Sun Zi'an, 1995.)

The danger of allowing an adversary, particularly one with the high-tech capabilities of the United States, to seize the strategic initiative in a conflict places a premium on China's ability to apply military force very quickly at the outset of a conflict. A *Liberation Army Daily* article notes,

> For the weaker party, waiting for the enemy to deliver the first blow will have disastrous consequences and may even put it in a passive situation from which it will never be able to get out. (Lu, 1996.)

A number of Chinese publications argue that a preemptive or surprise assault against enemy forces is a logical, perhaps even necessary, military option for China when confronting an adversary like the United States. They contend that preemptive strikes represent a crucial way for a weaker country to overcome a more powerful country's advantages in limited, high-tech war. The PLA, they argue, will be able to seize the strategic initiative by striking such targets as key C^3I systems, military bases, and logistic centers with long-range missiles and attack aircraft before the U.S. deployment is complete or defenses are established. (National Defense University, 1993, pp. 6–15.)

Engaging in a preemptive strike runs counter to the PLA's traditional doctrinal guideline of "striking only after the enemy has struck." The

---

[3]The phrase "fleet of aircraft carriers" may be a mistranslation for "carrier battle groups."

[4]This assumes the Chinese are able to solve the problem of detecting, tracking, and targeting U.S. ships and planes in real time.

article cited above appears to get around this restriction by defining the "first shot" against China in strategic terms as "all military activities conducted by the enemy and aimed at breaking up China territorially and violating its sovereignty." By this definition, any U.S. military support for Taiwan in the context of a military crisis with the mainland could be construed as an activity designed to break up China and as the equivalent of a "strategic first shot" by the United States against the PRC. Beijing could, therefore, legitimately launch an attack against U.S. forces within the "strategic framework of gaining mastery by striking only after the enemy has struck." (Lu, 1996.)

POLITICAL STRATEGIES

In the context of local warfare, the use of military force is far more subject to the demands of politics and diplomacy than in previous forms of warfare. Thus, attention must be paid to the political strategies that can complement the military one already discussed. The ability to dominate a situation politically can allow the effective use of force when the adversary has greater absolute military capabilities.

If a military conflict with the United States becomes likely, a clear objective of Beijing would be to undermine U.S. political support in Asia. While potentially difficult, isolating the United States from key allies in Asia would significantly increase the political and financial costs to the United States in conducting operations against China. This may be sufficient to dissuade or deter the United States from engaging in military conflict with China at all. As one article appearing in *Chinese Military Science* explained,

> It is proper to imagine that if there should occur a war crisis in the future, the probability of the United States authorities deciding to use forces will be greatly reduced if the United States can be divided from its key allies. (Fu, 1997.)

The most important U.S. ally in Asia is Japan. Chinese periodicals do not discuss pressuring Japan as a policy option. Nonetheless, doing so follows logically from the sentiment expressed above. Japan possesses Asia's most modern military and is home to U.S. military bases that would play a vital role in any U.S. military effort against China.

Japanese neutrality or imposition of restrictions on U.S. use of its bases in Asia would be of enormous benefit to China in the event of a military conflict with the United States. Moreover, China does have some leverage at its disposal. First, China is an important market for Japan, both in terms of trade (over $60 billion in 1996) and investment. Restricting Japan's access to that market could have a significant effect on Japan's economy. Secondly, China's ballistic missile capabilities are a credible military threat to the Japanese islands. While pressuring Japan economically or militarily could backfire and bring serious costs to China's own economy and security, such pressure could also convince Tokyo that its long-term interests will not be served by supporting the United States in a conflict with the PRC.

In the context of actual military conflict with the United States, China could also attempt to undermine political support for the conflict within the United States. As mentioned earlier, it appears to be a common view among Chinese observers of the United States that Americans have a strong aversion to heavy casualties. (Fu, 1997, and Yu, 1996.) As one PLA colonel put it, "Fearing death and casualties are [sic] their fatal weakness." (Yu, 1996.) Pointing to the U.S. experience in Somalia and Lebanon as evidence, Chinese strategists contend that there will be substantial public opposition to U.S. participation in any conflict in which significant casualties are incurred. As a result, U.S. military strategy places a priority on minimizing potential casualties and tries to avoid confronting nations that have demonstrated the willingness to make great sacrifices for their nations' interests. U.S. policymakers will hesitate to engage in conflict with a nation demonstrating the courage or capability "to fight a bloody war." (Fu, 1997.) In the context of rising tensions between China and the United States, Chinese leaders may conclude it to be in their interests to emphasize their country's willingness to accept high casualties and its ability to inflict them upon the United States.

Chapter Seven

CONCLUSIONS

The conclusion of this analysis is that China, despite an awareness of its relative weakness, might nevertheless be willing to use force against the United States or in a way that runs a major risk of U.S. involvement.[1] Among other implications, this poses an important challenge to intelligence analysis. The nature of this challenge is demonstrated by the Egyptian-Syrian attack on a stronger Israel in October 1973, the possibility of which was much underestimated on the grounds that the overall military balance suggested the Arab side could not possibly win a war against Israel.

In using force despite an overall unfavorable military balance, China would be primarily seeking to achieve a political effect. One can easily imagine circumstances in which the Chinese might believe it could obtain such a political effect. The most obvious cases would deal with Taiwan and would seek to exploit the ambiguities in the U.S. commitment to defend Taiwan, especially under circumstances in which a Taiwanese action (perhaps even a less-provocative action than a unilateral declaration of independence) could be said to have precipitated the crisis. In such a case, a Chinese use of force could be intended to affect U.S. policy and drive a wedge between Taiwan and the United States. For example, by posing an ambiguous threat to Taiwan, China might hope that differing perceptions of the degree of threat (less in the United States than in Taiwan) could lead to a sense

---

[1]Finkelstein (1999), p. 111, reports that the "president of a very prestigious Chinese think-tank" told him that "even when the correlation of military forces is obviously not in our favor China will still go to war over the issue of sovereignty" (which, of course, would include a Taiwan scenario).

in Taiwan that the United States was preparing to abandon it. In any case, China would wish to bring home to the U.S. public the potential risks of a pro-Taiwan policy.

Force could also be used to influence the political situation on Taiwan. By raising fears of a major military action, China might hope to exacerbate tensions on Taiwan between those willing to run major risks in the name of eventual independence and those who, whatever their abstract views about independence, did not wish to do so. In general, China might view Taiwan's political situation as fragile and might believe that a crisis could tilt the balance of the countervailing pressures in favor of accommodation with the mainland.[2]

Thus, the best indicator of possible Chinese action might be their estimate of the political situation, either in the United States or on Taiwan. Unfortunately, this poses a much more difficult problem for U.S. intelligence-collection methods than estimating the military balance of power.

Other possibilities for Chinese use of force are more remote. One could imagine Chinese military actions in the South China Sea in support of its territorial claims. While this could easily involve minor incidents (such as naval skirmishes or attacks on islands held by other claimants), it is more difficult to imagine how a major clash between the United States and China might result.

In any conflict or potential conflict with the United States, China, understanding that it is the generally weaker party, would have to look for asymmetric strategies that would provide leverage against the United States. As explained above, it would seek ways to exploit U.S. vulnerabilities and to prevent the United States from bringing its superior force to bear. Fundamentally, China would seek to create a *fait accompli*, thereby forcing the United States, if it wished to rein-state the *status quo ante*, to escalate the level of tension and violence. China would then count on the pressure of "world public opinion," the general disinclination to see profitable economic relationships

---

[2]Taiwanese nationalism and a desire to preserve the gains Taiwan has made in estab-lishing a democratic system of government would lead Taiwanese opinion to resist Chinese pressures for reunification; concern for Taiwanese business interests on the mainland, as well as a general fear of instability, would influence Taiwanese opinion in favor of concessions on the reunification issue.

disrupted, and on U.S. public opinion to constrain the United States in this situation.

In support of this goal, China would seek to suppress the U.S. ability to project substantial military resources into the theater of conflict for a limited time, either by information warfare attacks or by missile attacks (or the threat of them) on ports, airfields, transit points, bases, or other key facilities in the western Pacific. At least some Chinese strategists believe this can be accomplished through the use of ballistic or cruise missiles targeted at important but vulnerable "nodes" in U.S. command, control, communications, computers, intelligence, surveillance, and reconnaissance systems. China could implicitly or explicitly threaten to use chemical or biological weapons on such facilities; the result might be that facilities located in foreign countries were not made available for U.S. military use.

China could seek to cause U.S. casualties, to shock U.S. public opinion. The articles that discuss this issue frequently point to the U.S. experiences in such places as Somalia and Lebanon, where U.S. forces did withdraw after suffering unexpected casualties. In addition, such losses could lead the U.S. public to think that the policy its government was following might lead to a major war. A major psychological shock would be sought: China could seek to do something for which public opinion would not at all be prepared. For example, it could attempt to attack a carrier battle group; if it succeeded in causing major damage (especially to a carrier), the domestic political reaction could be intense.[3]

Finally, China could seek to exploit the fact that the United States was dealing with a crisis elsewhere in world; for example, it could act against Taiwan at a time when the United States had made major deployments to the Persian Gulf.

Needless to say, these are extremely risky strategies, since, like the Japanese attack on Pearl Harbor, they could easily lead to a major U.S. response that, in time, would prove overwhelming. The Chinese leadership could not embark on them unless it was confident that it could assess the likelihood of such a U.S. reaction. Among other

---

[3]This is obviously a tall order for the current Chinese military. However, if one assumes that the attack were to take place during a crisis, before the U.S. forces were on a war footing, it might be possible.

things, China would have to take into account the U.S. stake in the conflict and could not reasonably expect that losses of the magnitude that led to the U.S. pullout from Somalia would have the same effect in East Asia, although how the Chinese leadership might assess such a question is far from clear.

Similarly, the Chinese leadership would have to assess the risks of widening the conflict if it, e.g., threatened to attack U.S. bases or port or other facilities to which U.S. forces had access located on the territories of third-party countries. While Chinese attacks, or threats of attack, against such facilities could delay U.S. deployments via those facilities and/or complicate U.S. access to them, the result could also be to increase the hostility of third-party countries. China could find that its action had served mainly to strengthen the cohesion and will of a coalition directed against it.

The PRC's historical record shows that its leaders have been willing to take risks of this sort and have, in fact, been quite successful in assessing them. In the historical cases discussed earlier, the PRC has been able to modulate the risk to avoid a massive reaction from its stronger adversaries (both the United States and the former Soviet Union). In many cases, China tried to create the appearance of engaging in bolder actions than it was really undertaking; during the Taiwan Strait crises, for instance, it apparently relied on the United States to understand that, despite its rhetoric about "liberating Taiwan," it was in fact posing no such threat.

China's past success in assessing and modulating the risks it was running may give it confidence that it will be able to do so in the future as well. In addition, China may feel, now and in the future, that it can afford to accept greater risks. Many of the past uses of force occurred when China either was not a nuclear power or did not have a secure nuclear second strike capability. The possession of strategic nuclear weapons may enable the Chinese leadership to run risks that it otherwise could not.[4]

On the other hand, China ran its past risks when it could, to some extent, count on support of one superpower against the other. Even

---

[4]To some extent, the "people's war" strategy, which postulated China's invulnerability to ultimate capture and defeat, served the same purpose.

in 1969, before the Sino-U.S. *rapprochement*, China gained some benefit from the superpower rivalry; the Soviet Union had to fear that the United States would exploit any opportunities created by a major war between it and China. The U.S. status as sole superpower reduces China's maneuvering room, which helps explain the Chinese preference for multipolarity. While the Chinese typically argue that the international system is evolving in the direction of multipolarity, it is also clear that this tendency has not progressed very far. In this respect, then, China faces a more difficult environment in which to run risks.

A NOTE ON CHINESE STRATEGIC CULTURE

The issue of whether China's "strategic culture" influences its international behavior, in general, and its proclivity to resort to force in international disputes, in particular, has recently been the subject of a major work by Alastair Iain Johnston. (Johnston, 1995.) Johnston seeks to refute what he takes to be the "almost monolithic" view of the scholarly community that "the Chinese strategic tradition is uniquely antimilitarist." According to this view, Chinese tradition demonstrates a "preference for stratagem, minimal violence, and defensive wars of maneuver or attrition" and "stresses indirection and the manipulation of the enemy's perceptions of the structure of the conflict" as opposed to the concentration of "maximum momentum . . . at a decisive point." (Johnston, 1995, pp. 25–26.)[1]

By contrast, Johnston argues that the Chinese strategic culture embodies an essentially *realpolitik* view, which he refers to as the *parabellum* paradigm, after the Roman adage: *si vis pacem, para bellum* ("if you want peace, prepare for war"). In this regard, he claims, Chinese strategic culture does not differ significantly from typical Western views of the role of force in an international system characterized by anarchy and the absence of a supranational force able to control it.

---

[1]See Johnston (1995), p. 26, for a list of the scholarly works that he cites as reflecting the prevailing view.

THE CASE FOR CHINESE EXCEPTIONALISM

Those who argue that Chinese strategic culture varies greatly from that of the West, and is essentially antimilitary in important respects, point to various aspects of Chinese culture and history to make their case. Among the most important are the Confucian tradition, the doctrines of the classical military authors (of whom Sun Zi is best known), and certain elements of China's historical experience.[2]

Confucian Tradition

The mainstay of this argument is the importance and longevity of the Confucian tradition in China, according to which it is not force, but the virtue of the ruler and his punctiliousness in performing the prescribed rites that determine the strength of the state and its ability to defend itself. The notion that the moral force of the ruler is more important than his military capabilities seems to be deeply rooted in this tradition.

As Arthur Waldron has pointed out, this idealistic approach is not as impractical as it might seem, given the conditions that prevailed during large parts of Chinese history. This tradition was developed in a period during which the various Chinese rulers were involved in struggles among a multiplicity of culturally Chinese states. To the extent that a given ruler excelled in the virtues that were looked up to throughout the entire system, it made sense to argue that this could be a political force in and of itself. The history of the time includes many cases in which a talented individual, born in one state, rises to prominence as an advisor to the ruler of another; in the absence of competing nationalisms, one can imagine that popular and elite allegiances could readily shift from one ruler to another.

---

[2]This entire discussion is subject to the caveat that the current Chinese state inhabits a very different strategic environment than that in which the Chinese strategic culture, whatever it was, developed. Traditionally, China was at the center of its world, by far larger, wealthier, more powerful, and more advanced technologically than any state with which it came into contact. Not surprisingly, its geopolitical centrality was part of its self-image. This world was, of course, shattered for good 150 years ago; thus, while parts of the tradition retain their relevance for contemporary China (Mao, for example, was fond of quoting Sun Zi), one must be cautious in assessing which parts of the tradition should be used to explain or predict current Chinese behavior. See Swaine and Tellis (forthcoming) for a detailed discussion of the relevance of Chinese historical tradition to current policy.

Of course, what might make sense in the context of wars between culturally Chinese states might not work with respect to "barbarians," who would be more impressed with a Chinese ruler's "foreignness" than with his virtue. Even so, the superiority of Chinese culture to that of the surrounding peoples might be considered so obvious as to motivate their voluntary Sinification and willingness to come under virtuous and benevolent Chinese rule. At least, that was the theory; over the long run, the spread of Chinese culture from its original base in North China southward into what is now regarded by everyone as China suggests that it was not entirely without empirical basis.

Thus, this Confucian perspective—despite its idealistic trappings—may nonetheless have some importance for hardheaded analysts. As Waldron, who often takes a "hard line" position with respect to China on current policy issues, notes:

> Even today, China is simply too big, and its population too numerous, to be ruled by force alone. Contemplating the sheer numbers involved, Western strategists have regularly counseled against fighting "a land war in Asia," but it is less often recognized that Chinese leaders have always faced pretty much the same problem. An army of a size that is reliable is not large enough to coerce all of China; and an army large enough to coerce all of China will not be reliable. Not that force is irrelevant. Far from it. Still, the Confucian notion of a leadership based on virtue turns out to make a lot of sense under such conditions. (Waldron, 1997.)

Sun Zi

Sun Zi, as a military writer, is not directly part of the Confucian tradition. However, many of his precepts appear to imply a denigration of force and combat in favor of nonviolent methods of achieving victory.

Perhaps Sun Zi's most famous statement in this context is that "to win one hundred victories in one hundred battles is not the acme of skill. To subdue the enemy without fighting is the acme of skill." (Sun Zi, 3:3.) This has been interpreted to imply a preference for, in Johnston's words cited above, "indirection and the manipulation of the enemy's perceptions of the structure of the conflict."

Immediately following is another famous passage, which ranks four methods of achieving victory, from the most preferable to the least (Sun Zi, 3:4–7):

> Thus, what is of supreme importance in war is to attack the enemy's strategy.
>
> Next best is to disrupt his alliances.
>
> The next best is to attack his army.
>
> The worst policy is to attack cities. Attack cities only when there is no alternative.

Sun Zi does not explicitly describe the methods to be used in pursuing the first two, preferred, methods (attacking the enemy's strategy and disrupting his alliances), but it is easy to conclude that they involve diplomatic and psychological tactics rather than the use of force.

Historical Experience

Finally, proponents of this argument point to some patterns of Chinese history that appear consonant with this interpretation of the classic writings. Thus, it has been argued that China has typically preferred diplomatic measures (such as divide-and-rule strategies) to deal with the barbarian threats; often, it resorted to the use of measures, such as gifts, subsidies, and the ritual trappings of imperial prestige, to ensure the quiescence of the various peoples on its periphery.

Ralph Sawyer summarizes this view of the traditional Chinese disdain for military as opposed to "diplomatic" measures as a means of resolving national security problems as follows:

> Despite incessant barbarian incursions and major military threats throughout their history, . . . Imperial China was little inclined to pursue military solutions to external aggression. Ethnocentric rulers and ministers instead preferred to believe in the myth of cultural attraction, whereby their vastly superior Chinese civilization, founded upon virtue and reinforced by opulent material achievements, would simply overwhelm the hostile tendencies of the uncultured. Frequent gifts of the embellishments of civilized life,

coupled with music and women, would distract and enervate even the most warlike peoples. If they were unable to either overawe them into submission or bribe them into compliance, other mounted nomadic tribes could be employed against the trouble-makers, following the time-honored tradition of "using barbarian against barbarian." (Sawyer, 1996, p. 3.)

THE CASE AGAINST CHINESE EXCEPTIONALISM

Despite these arguments, variations of which have been widely accepted in the scholarly community, a strong case can be made against the idea of a peculiarly Chinese "strategic culture" of antimilitarism.

The Actual Impact of Confucian Idealism

It is of course difficult to assess the actual impact of the Confucian "idealism" described above. One could see much of it as an "ideology" that justified the internal political structure of the regime, without necessarily limiting the actual options available to the poli-cymakers. Johnston suggests several ways in which Confucianism might have served such a purpose for the Ming dynasty decision-makers whose actual strategies, he found, were consistent with a *parabellum* strategic culture.[3]

In any case, the Confucian rhetoric, while elevating nonviolent methods of statecraft, also tended to see those who refused to submit to "benevolent" Chinese rule as immoral, if not subhuman. This contemptuous attitude thwarted attempts to follow policies that took into account the actual interests of the barbarians and sought to accommodate them where feasible. Thus, according to Waldron, the traditional idealism, given voice by the scholar-officials trained in the Confucian classics, made it difficult for Ming dynasty officials to fol-low pragmatic policies of trade and diplomatic maneuver with

---

[3]For example, decisionmakers might have used Confucian concepts and terminology "to frame decisions made for other reasons or by other people and groups [e.g., military commanders] in order to justify their competence as strategists to themselves and the community." Similarly, Confucian language may have been used "to obscure the gap between the professed values of the group and actual behavior." (Johnston, 1995, pp. 251–252.)

respect to the nomadic peoples based on the steppes of China's northern border. Instead, there were impractical attempts to subdue the nomads and, finally, to seal off China from the nomads altogether. As a result China relied on force more than it would have had it followed less "Confucian" policies.[4]

Understanding Sun Zi

The famous adages of Sun Zi, referred to above, certainly demonstrate his understanding of the burden that warfare places on the state and his keen appreciation of the danger of overburdening the populace to the point that its loyalty is endangered and the state is weakened. Even victory in a long and difficult war is a mixed blessing: "For there has never been a protracted war from which a country has benefited." (Sun Zi, 2:7.)[5]

Nevertheless, it is probably incorrect to understand Sun Zi solely in terms of these aphorisms. First, and most massively, one must note that the bulk of the *Art of War* is taken up with discussion of actual combat. Thus, regardless of the status of the ideal of "not fighting and subduing the enemy," it would appear that Sun Zi clearly recognizes that often this will not be the case.

In any case, it is not clear that this ideal implies that combat is unnecessary. It is at least equally possible that what is meant is that correct precombat maneuvering can enable one, in effect, to secure the victory before combat begins, i.e., one has so structured the situation that a favorable outcome to the actual combat, once it takes place, is all but inevitable:

> Thus a victorious army wins its victories before seeking battle: an army destined to defeat fights in the hope of winning. (Sun Zi, 4:14.)

---

[4]This theme is developed at length in Waldron (1994), pp. 85–114.

[5]But this is not to be confused with soft-heartedness: Later in the work, Sun Zi enumerates five possible character traits of a general that constitute "serious faults" and that lead to "the ruin of the army"—being of a "compassionate nature" is one of them. "If [the general] is of a compassionate nature you can harass him." (Sun Zi, 8:22.)

Similarly, the preference for attacking an enemy's strategy or alliances as opposed to his army or cities is not equivalent to an injunction not to use force: It is perfectly compatible with the notion of using force in a way calculated to achieve the desired objectives. For example, if an enemy's strategy is to build up its military forces slowly, so as eventually to exert an overwhelming pressure, the instigation of a violent incident (such as a border clash) may be an effective way to attack that strategy: It forces the enemy to consider that it will not be able to achieve its goals as safely and cheaply as it had hoped.[6] The same argument may be made about attacking alliances. Using armed force to create an international crisis may be an effective way to attack an enemy's alliances: It forcefully brings home to the enemy's current or potential allies the possible costs of the alliance and may cause them to reconsider.[7]

Historical Evidence

Chinese history, as one would expect, if only from its length, contains examples of many approaches to security issues; not all of them hew to the Confucian line of substituting the ruler's virtue and benevolence for the use of force in dealing with external threats. Some have argued that Chinese behavior in this regard has followed a generally cyclical pattern, corresponding to the rise and fall of the various dynasties. At the beginning of a dynasty, once internal control had been secured, China would typically take military steps to assert (or reassert) control over strategically important border areas.[8]

More generally, it has been argued that the Confucian tradition constitutes only one strand of Chinese military thought and practice. The other derives from the nomads of the steppes of Central Asia and

---

[6]The March 2, 1969, border clash between China and the Soviet Union may be an example of this. The Chinese attacks on Indian forces in the border region in fall 1962 likewise represented an attack on the enemy's strategy, in this case, India's "forward policy."

[7]The Chinese shelling of the offshore islands of Jinmen and Mazu in 1954 was intended in part to "break up the collaboration between the United States and Chiang Kai-shek, and keep them from joining together militarily and politically . . . ." (From a Mao Zedong telegram to Zhou Enlai, July 23, 1954, as quoted in Zhang, Shu Guang, 1992, p. 193.)

[8]On this subject, see Swaine and Tellis (forthcoming).

Mongolia, who periodically threatened the agriculturally based Chinese state. This strand placed greater emphasis on military power and favored offensive and counteroffensive military postures (as opposed to a defensive posture, as most notably symbolized by the Great Wall). (Waldron, 1994, pp. 88–89, 95–97.)

In any case, empirical studies of Chinese use of force do not substantiate the argument that Chinese policy has been notably pacific.[9]

SUN ZI AND CLAUSEWITZ

As this brief review of the debate seems to indicate, the notion of an abiding Chinese cultural aversion to the use of force seems overstated; nevertheless, there may be some differences between typical Chinese and western perspectives concerning the use of force. To get at this question from a different angle, we next compare the views of Sun Zi and Clausewitz.[10]

The Role of Intelligence

Perhaps the most immediately apparent and most striking contrast between the two writers has to do with their views on intelligence. Many well-known passages in Sun Zi deal explicitly with the decisive importance of intelligence (which in Sun Zi's time, of course, primarily meant espionage); most succinctly,

> Now the reason the enlightened prince and the wise general conquer the enemy whenever they move and their achievements surpass those of ordinary men is foreknowledge.

> What is called "foreknowledge" cannot be elicited from spirits, nor from gods, nor by analogy with past events, nor from calculations. It must be obtained from men who know the enemy situation. (Sun Zi, 13:3–4.)

---

[9]Some of the historical evidence is presented in Johnston (1995), p. 27.

[10]A discussion of the extent to which Clausewitz can be used as the source of a "standard" western view of war is beyond the scope of this paper. For present purposes, he is useful, not merely because of his canonical status, but also because his views on many important issues contrast sharply with those of Sun Zi.

In many other passages, Sun Zi makes clear that intelligence (in the sense of knowledge of the enemy's situation) is one of the most important factors in achieving victory.   Hence, aside from the "benevolence" or care for his troops that secures their loyalty to him, knowledge (of both the enemy and oneself) is the general's most important asset. Since accurate knowledge is in principle available, a general's plans—even complicated ones—can be formulated with confidence that they can be implemented.

One of Clausewitz's best known concepts, on the other hand, is the "fog of war," i.e., the uncertainty that inevitably envelops the battle-field. Thus, accurate, useful intelligence, however desirable it might be in the abstract, is simply not to be expected in actual combat. Numerous passages testify to this skeptical view of intelligence:

> Many intelligence reports in war are contradictory; even more are false, and most are uncertain. . . . It is much worse for the novice if [the intelligence reports he receives do not contradict each other], and on the contrary one report tallies with another, confirms it, magnifies it, lends it color, till he has to make a quick decision— which is soon recognized to be mistaken, just as the reports turn out to be lies, exaggerations, errors, and so on. In short, most intelligence is false, . . . [Ominous intelligence reports] may soon, like waves, subside; but like waves they keep recurring, without apparent reason. The commander must trust his judgment and stand like a rock on which the waves break in vain.[11] (Clausewitz, I:7, 117.[12])

Therefore, according to Clausewitz, the true test of the commander is not his ability to collect and understand intelligence about the enemy or to make complicated plans that depend on precise information, but rather his ability to act forcefully and purposefully in the absence of such information.

The same distinction appears with respect to the general's knowledge of his own troops and what they can do (and are doing). Clausewitz emphasizes the pervasiveness of "friction," i.e., the

---

[11]If one cannot trust even mutually confirming intelligence reports, the situation is indeed bleak.

[12]Note that this and subsequent references to Clausewitz's *On War* follow the pattern of book number preceding the colon, followed by the chapter number, then, after a comma, the page number. We have used the Howard and Paret (1976) translation.

innumerable obstacles one encounters in trying to implement one's plans under battlefield conditions: "Everything in war is very simple, but the simplest thing is difficult." (Clausewitz, I:7, 119.) Sun Zi, on the other hand, posits that a knowledge of one's own forces is not only necessary but possible.

The Limits of Rationality

This point brings us to a somewhat broader, albeit vaguer, difference between the two thinkers. Clausewitz emphasizes that war, even though it is fundamentally a continuation of politics by other means, nevertheless has a *tendency* to "slip the leash," i.e., to tend, according to its own nature, toward an "absolute" level of violence. This is not only because the (intermediate) goal of subjecting the enemy to one's will in effect takes over from the political goal for which the war was fought in the first place; it also reflects the fact that war exists in the element of chance and engages basic human passions that are not fully under rational control:

> chance: the very last thing that war lacks. No other human activity is so continuously and universally bound up with chance. And through the element of chance, guesswork and luck come to play a great part in war. (Clausewitz, I:1, 85.)

> If war is an act of force, the emotions cannot fail to be involved. War may not spring from them, but they will still affect it to some degree, and the extent to which they do so will depend not on the level of civilization but on how important the conflicting interests are and on how long their conflict lasts. (Clausewitz, I:1, 76.)

By contrast, Sun Zi appears to believe that, despite its uncertainties and horrors, war can be kept under rational control. Guided by knowledge of the enemy and of himself, the true general stays in control of the battle:

> In the tumult and uproar the battle seems chaotic, but there is no disorder; the troops appear to be milling about in circles but cannot be defeated. (Sun Zi, 5:17.)

"Order and disorder," according to Sun Zi (5:19), "depend on organization." In principle, the general can maintain control of his army:

> Generally, management of many is the same as management of few. It is a matter of organization.
>
> And to control many is the same as to control few. This is a matter of formations and signals. (Sun Zi, 5:1–2.)

Another sign of this belief that a pragmatic rationalism can maintain control of the seeming chaos of war without engaging the emotions is that the ideal general is not motivated by a desire for honor or glory. Indeed, Sun Zi says that the ideal general, whose victories appear deceptively simple because he has so completely prepared favorable conditions for victory before the actual combat begins, will not win glory:

> And therefore the victories won by a master of war gain him neither reputation for wisdom nor merit for valor. (Sun Zi, 4:11.)

Clausewitz (I:1, 86) says that "In the whole range of human activities, war most closely resembles a game of cards." For Sun Zi, the more appropriate analogy would be chess or go.

Stratagem Versus Overwhelming Force

As a result of these differences, Sun Zi and Clausewitz provide very different visions of the ways in which wars are won. As Sun Zi states at the beginning of his work (1:17–20, 22–23, 26–27),

> All warfare is based on deception.
>
> Therefore, when capable, feign incapacity; when active, inactivity.
>
> When near, make it appear that you are far away; when far away, that your are near.
>
> Offer the enemy a bait to lure him; feign disorder and strike him. . . .
>
> Anger his general and confuse him.
>
> Pretend inferiority and encourage his arrogance. . . .
>
> Attack where he is unprepared; sally out when he does not expect you.
>
> These are the strategist's keys to victory. It is not possible to discuss them beforehand.

As these verses make clear, the point of deception is not merely to deny accurate information to the enemy but to induce him to act in ways that are beneficial to oneself. Consequently, the use of stratagem becomes a defining quality of good generalship.

Given his skeptical view of the value of intelligence, it is not surprising that Clausewitz is much less sanguine about the possible benefits of deception, since deception depends to a large extent on understanding the enemy's frame of mind, his beliefs about oneself, etc.,

> Yet, however much one longs to see opposing generals vie with one another in craft, cleverness, and cunning, the fact remains that these qualities do not figure prominently in the history of war. . . .
>
> plans and orders issued for appearances only, false reports designed to confuse the enemy, etc. . . . have as a rule so little strategic value that they are used only if a ready-made opportunity presents itself. They should not be considered as a significant independent field of action at the disposal of the commander. (Clausewitz, III:10, 202–203.)

Instead, Clausewitz espouses a very different approach to obtaining victory. The key concept is the "center of gravity" against which one's energies must be directed:

> [O]ne must keep the dominant characteristics of both belligerents in mind. Out of these characteristics a certain center of gravity develops, the hub of all power and movement, on which everything depends. That is the point against which all our energies should be directed. (Clausewitz, VIII:4, 595–596.)

This strategy is necessarily a bold one, since it involves attacking important targets, to the defense of which a large part of the enemy's strength will be dedicated:

> Not by taking things the easy way—using superior strength to filch some province, preferring the security of this minor conquest to great success—but by constantly seeking out the center of his power, by daring all to win all, will one really defeat the enemy. (Clausewitz, VIII:4, 596.)

Determining the center of gravity must be done on a case-by-case basis; however, judging from "general experience," Clausewitz opines as follows:

[T]he acts we consider most important for the defeat of the enemy
are the following:

1. Destruction of his army, if it is at all significant

2. Seizure of his capital if it is not only the center of administration
    but also that of social, professional, and political activity

3. Delivery of an effective blow against his principal ally if that ally
    is more powerful than he. (Clausewitz, VIII:4, 596.)

The destruction of the enemy's army and the seizure of his capital
are thus the typical Clausewitzian prescriptions, whereas for Sun Zi
(who prioritizes targets in the following order: the enemy's strategy;
his alliances; his army; and, last, his cities), they would be the less
preferable means of conducting a war.[13]

Bloodless Warfare

The following Clausewitzian blast may be taken as his reply to Sun
Zi's famous formula of "not fighting and subduing the enemy"
(although Clausewitz undoubtedly had more recent theorists in
mind):

> Kind-hearted people might of course think there was some inge-
> nious way to disarm or defeat an enemy without too much blood-
> shed, and might imagine this is the true goal of the art of war.
> Pleasant as it sounds, it is *a fallacy that must be exposed*: war is such
> a dangerous business that *the mistakes which come from kindness
> are the very worst.* (Clausewitz, I:1, 75; emphasis added.)

It is hardly accurate to view Sun Zi as "kind-hearted" in this sense:
Among other bits of evidence, one might note that Sun Zi lists a

---

[13]One might argue that Clausewitz is hardly unaware of the importance of the goals
that Sun Zi emphasizes. Indeed, he notes that

> It is possible to increase the likelihood of success without defeating the enemy's forces.
> I refer to operations that have *direct political repercussions*, that are designed in the first
> place to disrupt the opposing alliance, or to paralyze it, that gain us new allies, favor-
> ably affect the political scene, etc. (Clausewitz, I:2, 92; emphasis in original.)

However, this insight is peripheral for Clausewitz, but central for Sun Zi. (As an aside,
one might note that this passage illustrates the point, made above, that Sun Zi's preferred approaches—attacking the enemy's strategy and alliances—need not involve
only diplomatic or nonviolent means.)

"compassionate nature" as one of five possible "serious faults" of a general that lead to "the ruin of the army."[14] Nevertheless, he does put forward an ideal of bloodless warfare that Clausewitz finds not only wrong, but pernicious.

The reason appears to be that, for Clausewitz, regardless of how clever one's maneuvers and stratagems, the opponent always retains the option of seeking a decision through a major battle. Hence, if we seek victory through maneuver, we become, in a sense, hostages to our opponent's plans:

> If he were to seek the decision through a major battle, *his choice would force us against our will to do likewise.* Then the outcome of the battle would be decisive. . . . (Clausewitz, I:2, 98; emphasis in original.)

Furthermore,

> it is clear—other things . . . being equal—that we would be at an overall disadvantage, since our plans and resources had been in part intended to achieve other goals, whereas the enemy's were not. . . . If, therefore, one of the two commanders is resolved to seek a decision through major battles, he will have an excellent chance of success if he is certain that his opponent is pursuing a different policy. (Clausewitz, I:2, 98.)

In this sense, then, for Clausewitz, the ideal of bloodless warfare is not only utopian but dangerous. Sun Zi, by contrast, would appear to believe that this option of decision by battle can be effectively denied to the opponent.

CONCLUSION

This appendix merely touches the surface of the question of Chinese strategic culture. Nevertheless, two conclusions seem relatively clear. First, any putative Chinese "antimilitarism" does not imply an unwillingness, or even a strong reluctance, to use force. It may imply a lesser regard for the typically military virtues and a lesser apprecia-

---

[14]"If [the general] is of a compassionate nature you can harass him." (Sun Zi, 8:22.)

tion of military glory. It may also involve a greater abhorrence of attrition as a military strategy. But it is not pacifism, and it is not a rejection of the use of force in principle. Rather, it seems to stem from a belief that force can be used in an extremely focused and rational manner.

Second, China may have, other things being equal, a tendency toward using force in ways that strikes us as peculiarly "political." Indian Major General D. K. Palit, reflecting on India's inability to understand Chinese strategy in the 1962 border war, has said that

> Concepts, such as limited wars, superpower management of local conflicts, proxy wars and coercive diplomacy had not yet modified the perception that decision was the goal of war. *It was left to the Chinese to point the way to subtleties of contemporary political and strategic manoeuvres.*" (Palit, 1991, p. 281; emphasis added.)

Angell, Norman, *The Great Illusion*, New York: Putnam's Sons, 1912.

Baum, Richard, "The Road to Tiananmen," in Roderick MacFarquhar, ed., *The Politics of China: 1949–1989*, New York: Cambridge University Press, 1993.

Blasko, Dennis, Philip Klapakis, and John Corbett, "Training Tomorrow's PLA: A Mixed Bag of Tricks," *The China Quarterly*, Vol. 146, June 1996, pp. 488–524.

Brzezinski, Zbigniew, *Power and Principle: Memoirs of the National Security Advisor, 1977–1981*, New York: Farrar Straus Giroux, 1983.

Chanda, Nayan, *Brother Enemy: The War After the War*, San Diego Calif.: Harcourt Brace Jovanovich, 1986.

Chen Bingde, "Intensify Study of Military Theory to Ensure Quality of Army Building: Learning from Thought and Practice of the Core of the Three Generations of Party Leadership in Studying Military Theory," in *Zhongguo Junshi Kexue*, No. 3, August 20, 1997, pp. 49–56, appearing as "Nanjing Military Region Commander Chen Bingde on Military Theory, Reform," Foreign Broadcast Information Service—China, March 6, 1998.

Christensen, Thomas J., "China, the U.S.-Japan Alliance and the Security Dilemma in East Asia," *International Security*, Vol. 23, No. 4, Spring 1999.

_____, "Chinese Realpolitik," *Foreign Affairs*, Vol. 75, No. 5, September/October 1996a.

_____, *Useful Adversaries: Grand Strategy, Domestic Mobilization, and Sino-American Conflict, 1947–1958,* Princeton, N.J.: Princeton University Press, 1996b.

Clausewitz, Carl von (1780–1831), *On War,* trans. Michael Howard and Peter Paret, eds., Princeton, N.J.: Princeton University Press, 1976.

Cohen, Eliot A., and John Gooch, *Military Misfortunes: The Anatomy of Failure in War,* New York: Vintage Books, 1991.

Cowan, Edward, "Carter Calls for Quick Withdrawal by China in Message to Its Leaders," *New York Times,* February 28, 1979, p. A4.

Deocadiz, Cristina, "US to 'Aid' Manila in Event of Spratlys Attack," *Manila Business World* (Internet Version), August 6, 1998, FBIS World News Connection, FBIS-EAS-98-218, August 7, 1998.

Ding Henggao, "New Defense Science and Technology Strategy to Emphasize Technology Transfer to Civilian Use," *Zhongguo Junshi Kexue* [*Chinese Military Science*], No. 3, August 20, 1995, pp. 131–136. (Appears as "COSTIND Director Ding Henggao on Defense S&T," Foreign Broadcast Information Service—China, August 20, 1995.)

Finkelstein, David M., "China's National Military Strategy" in James C. Mulvenon and Richard H. Yang, eds., *The People's Liberation Army in the Information Age,* Santa Monica, Calif.: RAND, CF-145-CAPP/AF, 1999.

"First Flight for F-10 Paves Way for Production," *Jane's Defense Weekly,* May 27, 1998, p. 17.

Fu Liqun, "Several Basic Ideas in US Strategic Thinking," *Zhongguo Junshi Kexue,* No. 1, February 20, 1997, pp. 28–37, 149. (Appears as "Academy of Military Science Officer on US International Strategy," Foreign Broadcast Information Service—China, February 20, 1997.)

Garver, John, "China's Push Through the South China Sea: The Interaction of Bureaucratic and National Interests," *The China Quarterly,* Vol. 128, December 1992, pp. 999–1028.

_____, *Foreign Relations of the People's Republic of China,* Englewood Cliffs, N.J.: Prentice Hall, 1993.

Gittings, John, *Survey of the Sino-Soviet Dispute: A Commentary and Extracts from the Recent Polemics, 1963–1967*, London: Oxford University Press, 1968.

Godwin, Paul, "Changing Concepts of Doctrine, Strategy and Operations in the Chinese People's Liberation Army 1978–87," *China Quarterly*, No. 112, December 1987, pp. 572–590.

_____, "Chinese Military Strategy Revised: Local and Limited War," *The Annals of the American Academy of Political Science*, Vol. 519, January 1992.

_____, "From Continent to Periphery: PLA Doctrine, Strategy and Capabilities Towards 2000," *China Quarterly*, No. 146, June 1996, pp. 464–487.

Godwin, Paul, and John Schultz, "Arming the Dragon for the 21st Century: China's Defense Modernization Program," *Arms Control Today*, December 1993, p. 4.

Hao Yufan and Zhai Zhihai, "China's Decision to Enter the Korean War," *The China Quarterly*, Vol. 121, March 1990, pp. 113–114.

He Di, "The Evolution of the People's Republic of China's Policy toward the Offshore Islands," in Warren I. Cohen and Akira Iriye, eds., *The Great Powers in East Asia, 1953–1960*, New York: Columbia University Press, 1990.

Hong Heping and Tian Xia, "Head to the New Century," *Zhongguo Kongjun* [*Chinese Air Force*], No. 5, October 1, 1996, pp. 4–7. (Appears as "Developing Air Force Officers for Next Century," Foreign Broadcast Information Service—China 97-009, October 1, 1996.)

Hong Ye, "The Western Nations Have Started a Smokeless War," *Zhenli De Zhuiqiu*, Beijing, November 11, 1995. (Appears as "On West's 'Smokeless War' of Containment," WNC-FBIS, December 22, 1995.)

Huang Jialun, "Three-Point Thinking on Developing Combat Theory," *Jiefangjun Bao*, April 7, 1998. (Appears as "Paper on New PLA Combat Theory Development," WNC-FBIS, May 1, 1998.)

Huang Rifei, "Chinese Special Forces Match the World Powers," in *Renmin Ribao* [*Peoples' Daily*], Guangzhou South China News Supplement, July 30, 1999, p. 1. (Appears in FBIS-CHI-1999-0730.)

Huang Xing and Zuo Quandian, "Holding the Initiative in Our Hands in Conducting Operations, Giving Full Play to Our Own Advantages to Defeat Our Enemy: A Study of the Core Idea of the Operational Doctrine of the People's Liberation Army," *Zhongguo Junshi Kexue*, No. 4, November 20, 1996, pp. 49–56. (Appears as "Operational Doctrine for High-tech Conditions," WNC-FBIS, June 17, 1997.)

Huang Youfu, Zhang Bibo, and Zhang Song, "New Subjects of Study Brought About by Information Warfare—Summary of Army Command Academy Seminar on 'Confrontation of Command on Information Battlefield," *Jiefangjun Bao*, December 20, 1997, p. 6. (Appears as "PLA Academy on IW-Related Study," WNC-FBIS, December 23, 1997.)

Hung, Nguyen Manh, "The Sino-Vietnamese Conflict: Power Play Among Communist Neighbors," *Asian Survey*, Vol. XIX, No. 11, November 1979, pp. 1037–1052.

Jia Wenxian, Zheng Shouqi, and Guo Weimin, "Tentative Discussion of Special Principles of a Future Chinese Limited War," *Guofang Daxue Xuebao* [*National Defense University Journal*], No. 11, November 1, 1987, pp. 8–9. (Appears as "Principles of Future Limited War," Joint Publications Research Service CAR-88-037, July 12, 1988, pp. 47–48.)

Jiang Deqing and Chen Ying, "Ministry of Electronic Industry's Institute 38 Has New Breakthroughs in Research on Ground Air-Defense Radar Technology," *Anhui Ribao*, June 22, 1993, p. 3. (Appears as "Decoy System, Other Air-Defense Radar Breakthroughs Announced," Joint Publications Research Service-*CST*, July 27, 1993.)

Jiao Wu and Xiao Hui, "Modern Limited War Calls for Reform of Traditional Military Principles," *Guofang Daxue Xuebao* [*National Defense University Journal*], No. 11, November 1, 1987, pp. 10–11, 58. (Appears as "Reform of Traditions 'Necessary,'" Joint Publications Research Service, July 12, 1988, pp. 49–51.)

Jencks, Harlan, "China's 'Punitive' War on Vietnam," *Asian Survey*, Vol. 19, No. 8, August 1979, pp. 801–815.

_____, "People's War Under Modern Conditions: Wishful Thinking, National Suicide, or Effective Deterrent?" *The China Quarterly*, No. 98, June 1984.

Joffe, Ellis, "People's War Under Modern Conditions: A Doctrine for Modern War," *The China Quarterly*, No. 112, December 1987, pp. 555–556.

Johnston, Alastair Iain, "China's New 'Old Thinking': The Concept of Limited Deterrence," *International Security*, Vol. 20, No. 3, Winter 1995/96.

_____, *Cultural Realism: Strategic Culture and Grand Strategy in Chinese History*, Princeton, N.J.: Princeton University Press, 1995.

Khalilzad, Zalmay M., Abram N. Shulsky, Daniel L. Byman, Roger Cliff, David T. Orletsky, David Shlapak, and Ashley J. Tellis, *The United States and a Rising China: Strategic and Military Implications*, Santa Monica, Calif.: RAND, MR-1082-AF, 1999.

Kennan, George F., *Memoirs: 1950–1963*, Vol. II, Boston: Little, Brown and Company, 1972.

Li Bingyuan, "Historical Mission of Soldiers Straddling 21st Century—Roundup of 'Forum for Experts on Meeting Challenge of the World Military Revolution,'" *Jiefangjun Bao*, January 2, 1996, p. 6. (Appears as "Forum Views Challenge of Military Revolution," WNC-FBIS, March 29, 1996).

Li, Nan, "The PLA's Warfighting Doctrine, Strategy, and Tactics," *The China Quarterly*, No. 146, June 1996.

"Li Peng's NPC Report: Military Cuts on Schedule," *Xinhua*, March 4, 1998. (Appears in WNC-CHI, March 8, 1998.)

Li Weixing, "Strengthening National Defense Building Is a Basic Guarantee for National Security and Economic Development—Thoughts on Studying Jiang Zemin's 'Correctly Handling Several Major Relationships in China's Socialist Modernization Drive,'" *Renmin Ribao*, March 7, 1996, p. 9. (Appears as "Relationship Between Defense, Economy Noted," Foreign Broadcast Information Service—China, March 7, 1996.)

Li Yaqiang, "Will Large-Scale Naval Warfare Recur?" *Jianchuan Zhishi*, August 8, 1995. (Appears as "Navy Officer on Likelihood of Large-Scale Naval War," Foreign Broadcast Information Service—China, August 8, 1996.)

Li Xiannian, "Warnings from Peking: Interview with PRC Deputy Prime Minister Li Xiannian," *Newsweek*, July 16, 1979, p. 59.

Liao Xilong, Commander of the Chengdu Military Region, "Readjust Thinking About Work: Make the Four Transformations as Fast as Possible," *Zhongguo Minbing* [*Chinese Militia*], December 9, 1997. (Appears as "Chengdu Military Region Commander on Mobilization, Technology" WNC-FBIS, March 1, 1998.)

Lieberthal, Kenneth, "The Great Leap Forward and the Split in the Yan'an Leadership, 1958–65," in Roderick MacFarquhar, ed., *The Politics of China*, New York: Cambridge University Press, 1993.

Lin Dong, "Applications of Digitization in Military," *Guofang*, No. 11, November 15, 1995, pp. 10–11. (Appears as "Academic Discusses Digitalization in PRC Military," Foreign Broadcast Information Service—China, November 15, 1995.)

Liu Fengcheng, "Concentrate Forces in New Ways in Modern Warfare," *Jiefangjun Bao*, November 21, 1995, p. 6. (Appears as "PRC New Forms of Concentration of Forces in Modern Warfare," WNC-FBIS, January 29, 1996.)

Liu Huaqing, "Unswervingly Advance Along the Road of Building a Modern Army with Chinese Characteristics," *Jiefangjun Bao*, August 6, 1993, pp. 1–2. (Appears as "Liu Huaqing Writes on Military Modernization," Foreign Broadcasting Information Service—China (hereafter Foreign Broadcast Information Service—China), August 18, 1993, pp. 15–22.)

Lu Linzhi, "Preemptive Strikes Crucial in Limited High-Tech Wars," *Jiefangjun Bao*, February 14, 1996, p. 6. (Appears as "Preemptive Strikes Endorsed for Limited High-Tech War," WNC-FBIS, February 7, 1996.)

MacFarquhar, Roderick, Timothy Cheek, and Eugene Wu, eds., *The Secret Speeches of Chairman Mao: From the Hundred Flowers to the Great Leap Forward*, Cambridge, Mass.: Harvard University Press, 1989.

Mao Zedong, *Selected Works of Mao Zedong*, Vol. I, Beijing: Foreign Language Press, 1975a.

_____, *Selected Works of Mao Zedong*, Vol. II, Beijing: Foreign Language Press, 1975b.

Maxwell, Neville, *India's China War*, New York: Random House, 1970.

"Media Report: Cross-Regional Training," FBIS-TRENDS, August 16, 1995.

Mulvenon, James, "The Limits of Coercive Diplomacy: The 1979 Sino-Vietnamese Border War," *Journal of Northeast Asian Studies*, Fall 1995, pp. 68–88.

Mulvenon, James, "The PLA and Information Warfare," in Mulvenon and Yang, 1999, pp. 175–186.

Mulvenon, James C., and Richard H. Yang, eds., *The People's Liberation Army in the Information Age*, Santa Monica, Calif.: RAND, CF-145-CAPP/AF, 1999.

Munro, Ross, "Eavesdropping on the Chinese Military: Where it Expects War—Where it Doesn't," *ORBIS*, Summer 1994, pp. 335–372.

National Defense University Science and Research Department, *Gao Jishu Jubo Zhangzheng Yu Zhanyi Zhanfa*, Beijing: Guofang Daxue Chuban [National Defense University Publication], 1993.

Naughton, Barry, "The Third Front: Defence Industrialization in the Chinese Interior," *The China Quarterly*, September, 1998, pp. 351–386.

Office of the Under Secretary of Defense for Acquisition & Technology, *The Militarily Critical Technologies List*, Part I: *Weapons Systems Technologies*, Washington, DC: U.S. Government Printing Office, August 1, 1996.

Palit, D. K., *War in High Himalaya: The Indian Army in Crisis, 1962*, New York: St. Martin's Press, 1991.

Pan Shiying, "Thoughts About the Principal Contradiction in Our Country's National Defense Construction," *Jiefangjun Bao*, April 14, 1987, p. 3. (Appears as "PLA Paper Views Defense Construction Contradiction," *Foreign Broadcast Information Service—China*, April 29, 1987, pp. K13–15.)

Periscope, *Daily Defense News* (Internet Edition), January 20, 1999.

Pillsbury, Michael, *Dangerous Chinese Misperceptions: Implications for DoD*, Washington, D.C.: Office of Net Assessment, 1998.

Robinson, Thomas W., *The Sino-Soviet Border Dispute: Background, Development and the March 1969 Clashes*, RM-6171-PR, Santa Monica, Calif.: RAND, 1970.

Ross, Robert S., *The Indochina Tangle: China's Vietnam Policy, 1975–1979*, New York: Columbia University Press, 1988.

Sawyer, Ralph, ed. and tr., *The Art of the Warrior: Leadership and Strategy from the Chinese Military Classics*, Cambridge, Mass.: Shambhala Publications, 1996.

Segal, Gerald, *Defending China*, Oxford: Oxford University Press, 1985.

Shambaugh, David, "China's Military: Real or Paper Tiger?" *Washington Quarterly*, Vol. 19, Spring 1996.

Shen Zhongchang, Zhou Xinsheng, and Zhan Haiying, "A Rudimentary Exploration of 21st Century Naval Warfare," *Zhongguo Junshi Kexue*, February 20, 1995. (Appears as "AMS Journal on 21st Century Naval Warfare," Foreign Broadcast Information Service—China.)

Shi Yukun, "Lt. Gen. Li Jijun Answers Questions on Nuclear Deterrence, Nation-state, and Information Age," *Zhongguo Junshi Kexue* [*China Military Science*], No. 3, August 20, 1995, pp. 70–76. (Appears as "General Li Jijun Answers Military Questions," Foreign Broadcast Information Service—China, August 20, 1995.)

Shi Zhongcai, "An Elementary Discussion of the Maturity of Modern Military Personnel," *Guofang* [*National Defense*], No. 5, May 15, 1996, pp. 11–13. (Appears as "Discussion on Developing Personnel for Modern War," Foreign Broadcast Information Service—China, May 15, 1996.)

"Shift in China's Strategy for Building Armed Forces Shows New Evaluation of World Situation," *Ta Kung Pao*, February 16, 1986. (Appears as "Armed Forces Strategy Shift Seen 'Revolutionary,'"

Foreign Broadcast Information Service—China, February 18, 1986, pp. W11–W12.)

"Statement by the Spokesman of the Chinese Government—a comment on the Soviet Government's statement of August 21," *Renmin Ribao* [*People's Daily*], September 1, 1963, as quoted in John Gittings (1968), p. 92.

Sun Zi (Sun Tzu), *The Art of War*, trans. Samuel Griffith, London: Oxford University Press, 1971.

Sun Zi'an, "Strategies to Minimize High-Tech Edge of Enemy," *Xiandai Bingqi* [*Modern Weaponry*], No. 8, August 8, 1995, pp. 10–11. (Appears as "Military Journal on Countering High-Tech Enemy," Foreign Broadcast Information Service—China, March 6, 1996, pp. 63–65.)

Swaine, Michael D., and Ashley J. Tellis, *Interpreting China's Grand Strategy: Past, Present, and Future*, Santa Monica, Calif.: RAND, MR-1121-AF, forthcoming.

Tang Guanghui, "An Analysis of Post–Cold War Security," *Shijie Zhishi* [*World Affairs*], No. 19, October 1, 1996, pp. 8–9, in Foreign Broadcast Information Service—China 97-029.

Tyler, Patrick E., "As China Threatens Taiwan, It Makes Sure U.S. Listens," *New York Times*, January 24, 1996, p. A3.

Valencia, Mark J., *China and the South China Sea Disputes*, Adelphi Paper 298, Oxford: Oxford University Press for IISS, 1995.

Waldron, Arthur, "The Art of Shi," *The New Republic*, June 23, 1997, p. 38.

_____, "Chinese Strategy from the Fourteenth to the Seventeenth Centuries," in Williamson Murray, MacGregor Knox, and Alvin Bernstein, eds., *The Making of Strategy: Rulers, States, and War*, New York: Cambridge University Press, 1994.

Wang Chunyin, "Characteristics of Strategic Initiative in High-Tech Local Wars," *Hsien-Tai Chun-Shih* [*CONMILT*], No. 245, June 11, 1997, pp. 19–22. (Appears as "Strategic Initiative in High-Tech Local Wars," WNC-FBIS, October 30, 1997.)

Wang Jianhua, "Weakness of Joint Military Operations," *Jiefangjun Bao*, April 15, 1997, p. 6. (Appears as "Joint U.S. Military Operation Assessed," WNC-FBIS, May 5, 1997.)

Wang Juanghuai and Lin Dong, "Viewing Our Army's Quality Building from the Perspective of What Information Warfare Demands," *Jiefangjun Bao*, March 3, 1998, p. 6. (Appears as "Implications of Information Warfare on PLA," WNC-FBIS, March 16, 1998.)

Wang Mingjin, "Guangzhou Theater Organizes Sea-Crossing Landing Campaign Exercises to Control the Enemy in the Air, Destroy the Enemy in the Sea, and Win Victory on the Island," *Jiefangjun Bao*, October 24, 1996, p. 1. (Appears as "Report on Guangzhou Theater Exercise of 'Seizing Islands,'" WNC-FBIS, October 30, 1996.)

Wang Xusheng, Su Jinhai, and Zhang Hong, "Information Revolution, Defense Security," *Jisuanji Shijie [China Computerworld]*, No. 30, August 11, 1997, p. 21. (Appears as "Information Revolution, Defense Security, WNC-FBIS, November 25, 1997.)

Xu Xiangqian, "Strive to Achieve Modernization in National Defense—in Celebration of the 30th Anniversary of the Founding of the People's Republic of China, *Hongqi*, No. 10, October 2, 1979, pp. 28–32. Appears as "*Hongqi* Carries Xu Xiangqian Article on National Defense," Foreign Broadcast Information Service—China, October 18, 1979, pp. L12–L19.

Yan Youqiang and Chen Rongxing, "On Maritime Strategy and the Marine Environment," *Zhongguo Junshi Kexue*, No.2, May 20, 1997, pp. 81–92. Appears as "Naval Officers on International, Chinese Maritime Strategy," WNC-FBIS, October 14, 1997.)

Yang Dezhi, "A Strategic Decision on Strengthening the Building of Our Army in the New Period," *Hongqi*, No. 15, August 1, 1985, pp. 3–7. Appears as "Yang Dezhi Writes Hongqi Article on Army Reform," Foreign Broadcast Information Service—China, August 8, 1985, p. K3.

Yu Guohua, "On Turning Strong Force Into Weak and Vice-Versa in a High-Tech Local War," *Zhongguo Junshi Kexue*, No. 2, May 20, 1996, pp. 100–104. (Appears as "NDU Officer on Weaker Force Achieving Victory in Local War," Foreign Broadcast Information Service—China, May 20, 1996.)

Zhang Dejiu, "In-Depth Information Warfare is Philosophical Warfare (Excerpt of Maj. Gen. Xu Yanbin's Academic Report at National Defense University)," *Jiefangjun Bao*, August 13, 1996, p. 6. (Appears as "Xu Yanbin Discusses Information Warfare," Foreign Broadcast Information Service—China, August 13, 1996.)

Zhang Dezhen and Zhu Manting, "Yearender: Relations Among Big Nations Profoundly Adjusted and Multipolar Trends Quickened," *Renmin Ribao*, December 15, 1997, p. 7. (Appears in FBIS-CHI-98-001.)

Zhang Guoyu, "Work Hard to Create a New Pattern for Methods of Operation: Roundup of Experiences in All-Army Summarization and Demonstration of Results in Studies on Methods of Operation," *Jiefangjun Bao*, October 24, 1995 p. 6. (Appears as "Regions Study Methods of Operation," Foreign Broadcast Information Service—China, October 24, 1995.)

Zhang, Shu Guang, *Deterrence and Strategic Culture: Chinese-American Confrontations, 1949–1958*, Ithaca, N.Y.: Cornell University Press, 1992.

Zhang Xiaojun, "Modern National Defense Needs Modern Signal Troops," *Guofang*, No. 10, October 15, 1995, pp. 20–21. (Appears as "Modern Military Communications Deemed Essential," Foreign Broadcast Information Service—China, October 15, 1995.)

Zhao Nanqi, "On Deng Xiaoping's Thinking on Military Reform," *Zhongguo Junshi Kexue*, No. 4, November 20, 1996, pp. 12–21. (Appears as "Gen. Zhao Nanqi on Deng Xiaoping's Military Reform," Foreign Broadcast Information Service—China, November 20, 1996.)

Zhao Ying, "A General Discussion of National Economic Security," *Shijie Zhishi* [*World Affairs*], No. 20, October 16, 1996, pp. 4–5. (Appears in Foreign Broadcast Information Service—China 97-017.)

Zheng Shenxia and Zhang Changzhi, "Air Power as Centerpiece of Modern Strategy," *Zhongguo Junshi Kexue*, No. 2, May 20, 1996, pp. 82–89. (Appears as "Officers View Impact of Growing Air Power Strategy," Foreign Broadcast Information Service—China, May 20, 1996.)

Zhi Chongduo and Liang Yongchang, "On Information Offensive, Defensive," *Jiefangjun Bao*, August 27, 1996, p. 6. (Appears as "Military Forum on Information Offensive, Defense," WNC-FBIS, October 9, 1996.)

Zong He, "Tentative Discussion on the Characteristics of Modern Warfare," *Shijie Zhishi* [*World Knowledge*], No. 15, August 1, 1983, pp. 2–4. (Appears as "PRC Journal Details Traits of Modern Warfare," Joint Publications Research Service, October 11, 1983, p. 78–83.)